555 Timer Applications Sourcebook, With Experiments

by

Howard M. Berlin

Originally published as
The 555 Timer Applications Sourcebook, with Experiments
by E & L Instruments, Inc.

Howard W. Sams & Co., Inc.
4300 WEST 62ND ST. INDIANAPOLIS, INDIANA 46268 USA

FIRST EDITION
THIRD PRINTING—1980

International Standard Book Number: 0-672-21538-1
Library of Congress Catalog Card Number: 78-56584

Printed in the United States of America.

Preface

Approximately five years ago, a new and revolutionary type of linear integrated circuit was developed. The 555 timer, or "The IC Time Machine," as it has been called, provided circuit designers and amateur experimenters with a relatively inexpensive, stable, and easy-to-use integrated circuit for both monostable and astable applications. Since this device was first made commercially available, a myriad of novel and unique circuits have been developed and presented in several trade, professional, and hobby publications.

This book is about the 555 timer. It will show you what the 555 timer is and how to use it, by itself, and with other solid state devices, without having to become an electrical engineer. It was primarily written to fulfill the many requests of a co-worker who would always come into my office and say: "Hey, do you have a 555 circuit that will do . . . ?" This was because I had a folder, about one inch thick, that was full of such circuits. In addition to this book being a sourcebook of the many circuits and techniques in use, perhaps it will encourage you to discover other interesting applications for the 555 timer.

In Chapter 1, we look at the timer's organization and electrical characteristics. Although the "guts" of the 555 timer contain a maze of transistors, resistors, and diodes, the internal organizations will be simplified. Then, in Chapter 2, we begin learning how to connect this device as a monostable multivibrator, or one-shot pulser. Chapter 3 describes the useful free-running astable multivibrator square-wave generator, without the need for expensive crystals. Next, Chapter 4 illustrates the uses of the 555 timer to form regulated positive and negative power supplies, or dc-dc converters, and switching regulators.

Many useful and novel circuits for the measurement and control of electrical parameters are presented in Chapter 5, which should benefit the amateur experimenter who is on a budget. The next four chapters are devoted to other specialized applications, including electronic games, telephone, music, and automotive circuits, as well as circuits for the home, the photographer, and ham and CB radio.

Chapter 10 allows you to gain a basic practical knowledge using the 555 timer. There are a total of 17 simple experiments, which you can breadboard in a few minutes, that are designed to demonstrate many of the timer's features and applications. Although many of the components for these experiments may already be on hand, an experimenter's kit is available from E&L Instruments, Inc.

While reading this book, you will hopefully be convinced that the 555 timer is almost as versatile as the operational amplifier. Almost every application and technique is presented, illustrating over 100 published and unpublished circuits, graphs, and tables, the majority of which have been personally wired and tested.

I would like to first acknowledge the assistance given me by Mr. Kirkman Phelps, and Mr. Curtis Bauer of Edgewood Arsenal. In addition, I would like to thank authors Mr. David G. Larsen, Dr. Peter R. Rony, and Mr. Jonathan A. Titus, for their encouragement and assistance. I am also grateful to Ms. Nancy Hotter, a technical writer intern at Edgewood Arsenal, for her careful review of this manuscript. Although an English major with no prior experience with electronics, she nevertheless was able to provide many helpful suggestions, especially after she performed most of the experiments presented in the Blacksburg Continuing Education Series books entitled *Logic & Memory Experiments Using TTL Integrated Circuits*.

I am also indebted to the publishers of the many trade and hobby magazines for allowing me to reproduce many of the illustrations presented in this book. But most of all, I want to thank my wife Judy for her patience and understanding while I spent many long evenings and weekends in the preparation of this book.

HOWARD M. BERLIN

This book is dedicated to my father

Contents

CHAPTER 6

CHAPTER 7

CHAPTER 8

CHAPTER 9

CHAPTER 10

APPENDIX

CHAPTER 1

Introduction to the 555 Timer

Before we can fully appreciate the many applications that can be used with the type 555 timer presented in this book, we should take a look inside the device and see what it does.

The 555 integrated-circuit timer is a monolithic timing circuit packaged either as an 8-pin circular style TO-99 can, as an 8-pin mini-DIP, or as a 14-pin DIP, as shown in Fig. 1-1. Although the Signetics Corporation first introduced this device as the SE555/NE555, other manufacturers have since produced and marketed their own versions, which are summarized in Table 1-1. Recently, several companies have packed two independent 555 timers into a 14-pin DIP unit, called the 556 dual timer, shown in Fig. 1-2. An exception is the D555, for dual 555, which is made by Teledyne. As seen in Table 1-1, most of the manufacturers make two types of 555 timers. In this case, the first number represents the type preferred for

Table 1-1. 555 Timer Manufacturers

Manufacturer	Type Number
Exar	XR-555
Fairchild	NE555
Intersil	SE555/NE555
Lithic Systems	LC555
Motorola	MC1455/MC1555
National	LM555/LM555C
Raytheon	RM555/RC555
RCA	CA555/CA555C
Texas Instruments	SN52555/SN72555

military applications which have somewhat improved electrical and thermal characteristics over their commercial counterparts. This is analogous to the 5400 and 7400 series convention for TTL integrated circuits.

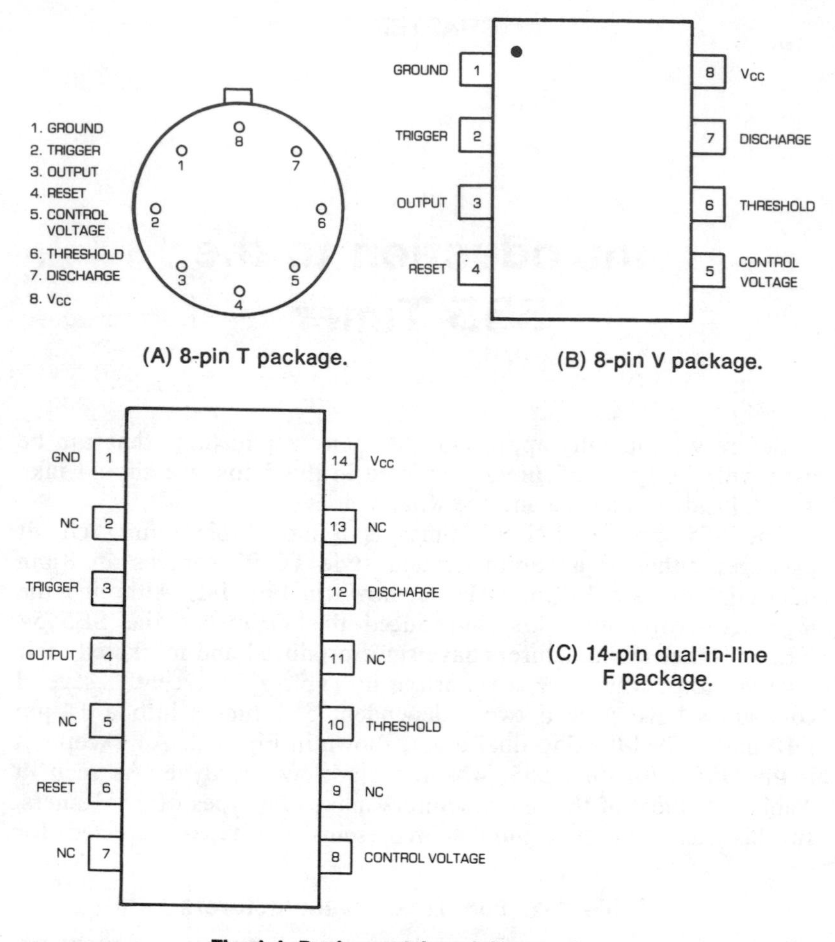

Fig. 1-1. Package styles of the 555 timer.

Inside the 555 timer is the equivalent of over 20 transistors, 15 resistors, and 2 diodes, depending on the manufacturer. The equivalent circuit for the 555 timer made by Signetics is shown in Fig. 1-3, and for comparison, the devices made by RCA, National Semiconductor, and Exar are shown in Figs. 1-4, 1-5, and 1-6 respectively. In any case, any of these equivalent circuits can be simplified to the block diagram of Fig. 1-7, providing the functions of control, trigger-

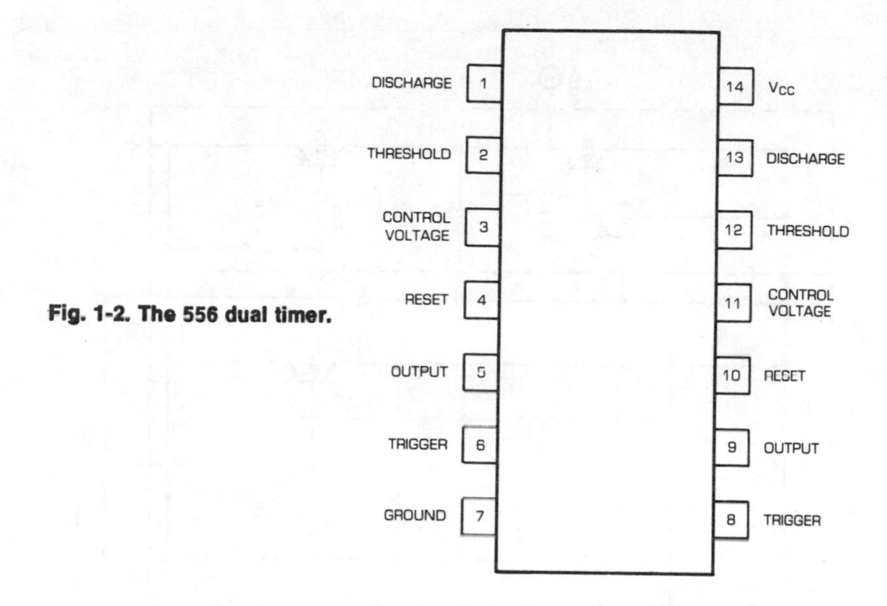

Fig. 1-2. The 556 dual timer.

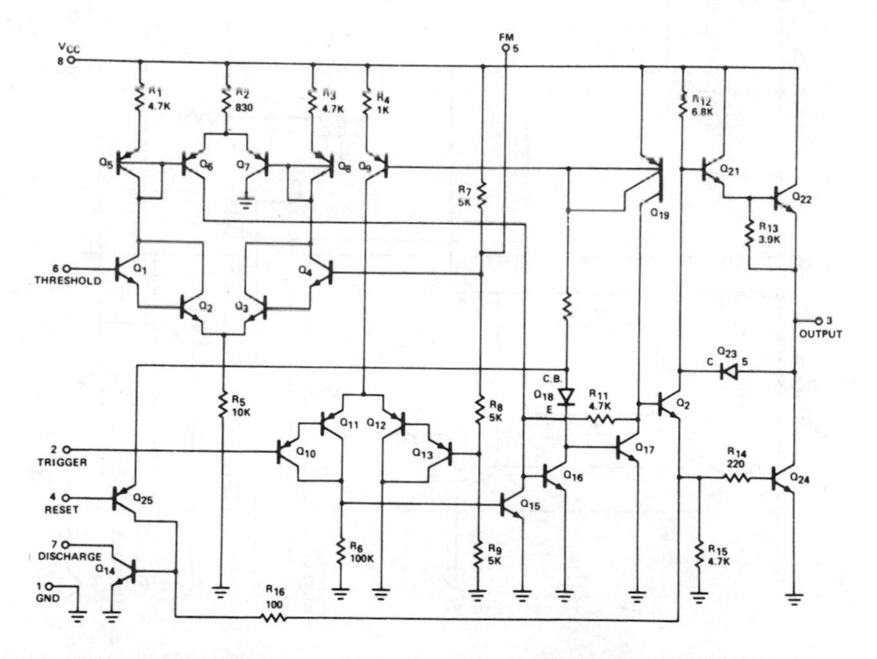

Courtesy Signetics, Sunnyvale, CA.

Fig. 1-3. 555 timer equivalent circuit.

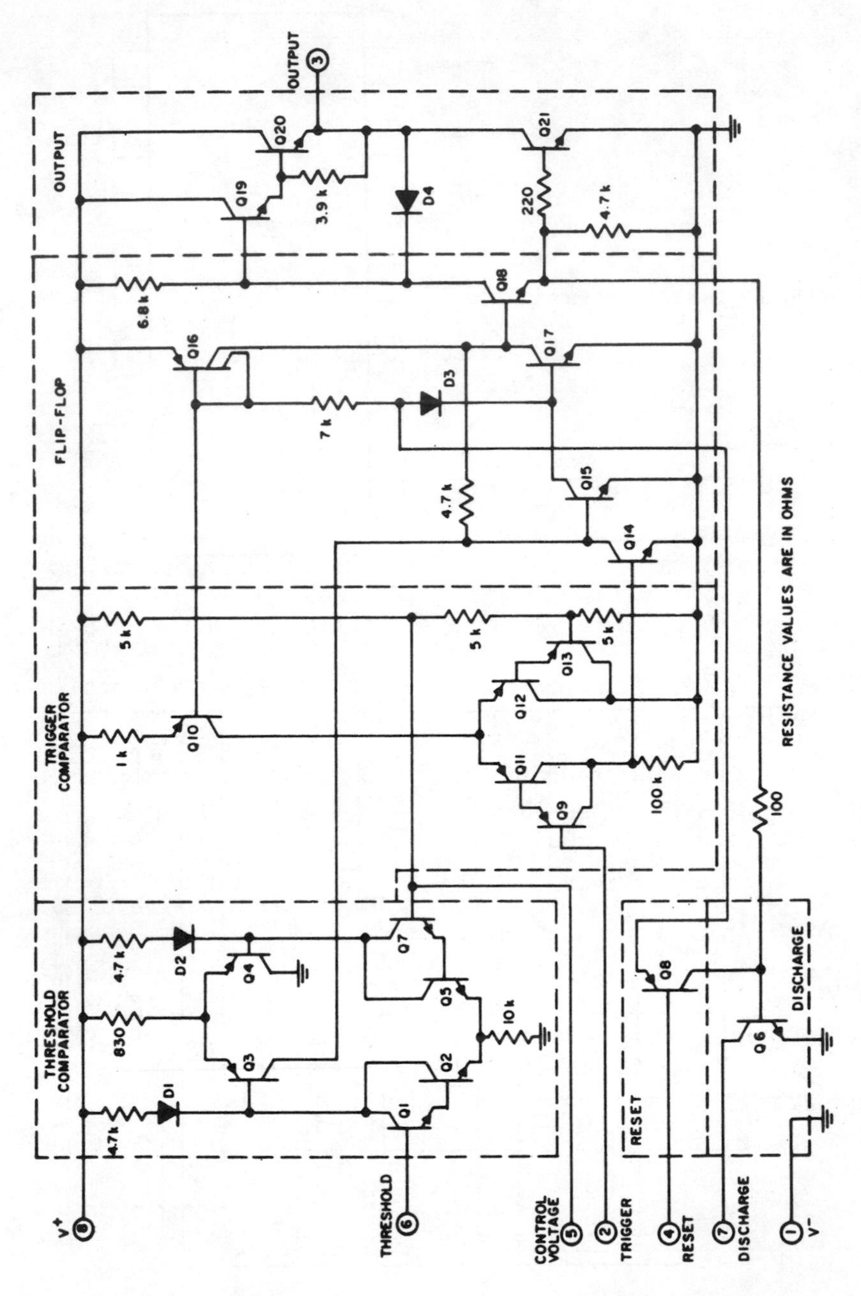

Courtesy RCA, Somerville, NJ.

Fig. 1-4. 555 timer equivalent circuit.

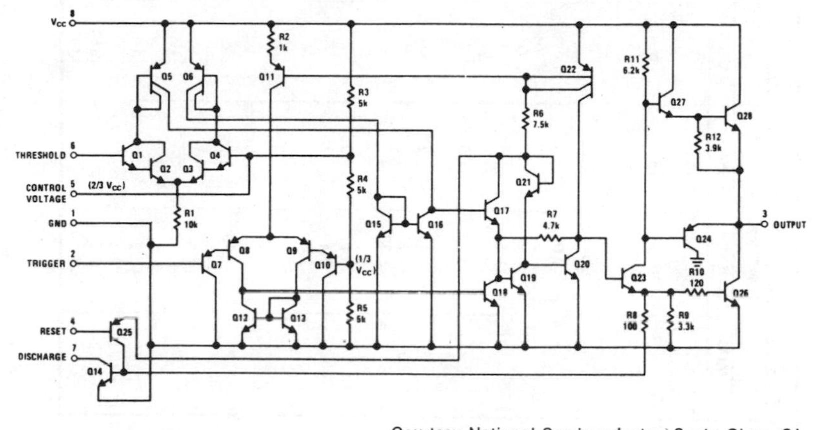

Courtesy National Semiconductor, Santa Clara, CA.

Fig. 1-5. 555 timer equivalent circuit.

ing, level sensing or comparison, discharge, and power output. In Chapters 2 and 3, the functions of each stage as they relate to monostable and astable operation will be explained.

Table 1-2 lists the specific electrical characteristics for the 555 timer. These are typical for all of the types listed in Table 1-1. Fig. 1-8 shows the typical performance curves. We should note that the 555 timer possesses a high degree of accuracy and stability. The initial monostable timing accuracy is typically within 1% of its calculated value, and exhibits negligible (0.1%/V) drift with the supply

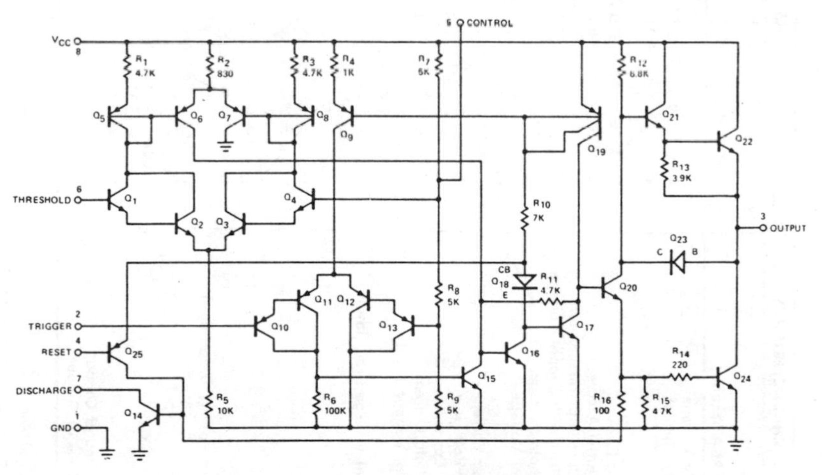

Courtesy Exar Integrated Systems, Sunnyvale, CA.

Fig. 1-6. 555 timer equivalent circuit.

Table 1-2. The 555 Linear Integrated Circuit

ELECTRICAL CHARACTERISTICS ($T_A = 25°C$, $V_{CC} = +5V$ to $+15$ unless otherwise specified)

PARAMETER	TEST CONDITIONS	SE 555			NE 555			UNITS
		MIN	TYP	MAX	MIN	TYP	MAX	
Supply Voltage		4.5		18	4.5		16	V
Supply Current	$V_{CC} = 5$ V $R_L = \infty$		3	5		3	6	mA
	$V_{CC} = 15$ V $R_L = \infty$		10	12		10	15	mA
	Low State, Note 1							
Timing Error	R_A, $R_B = 1$ KΩ to 100 KΩ							
Initial Accuracy	$C = 0.1$ µF Note 2		0.5	2		1		%
Drift with Temperature			30	100		50		ppm/°C
Drift with Supply Voltage			0.005	0.02		0.01		%/Volt
Threshold Voltage			2/3			2/3		X V_{CC}
Trigger Voltage	$V_{CC} = 15$ V	4.8	5	5.2		5		V
	$V_{CC} = 5$ V	1.45	1.67	1.9		1.67		V
Trigger Current			0.5			0.5		µA
Reset Voltage		0.4	0.7	1.0	0.4	0.7	1.0	V
Reset Current			0.1			0.1		mA
Threshold Current	Note 3		0.1	.25		0.1	.25	µA
Control Voltage Level	$V_{CC} = 15$ V	9.6	10	10.4	9.0	10	11	V
	$V_{CC} = 5$ V	2.9	3.33	3.8	2.6	3.33	4	V
Output Voltage Drop (low)	$V_{CC} = 15$ V							
	$I_{SINK} = 10$ mA		0.1	0.15		0.1	.25	V
	$I_{SINK} = 50$ mA		0.4	0.5		0.4	.75	V
	$I_{SINK} = 100$ mA		2.0	2.2		2.0	2.5	V
	$I_{SINK} = 200$ mA		2.5			2.5		
	$V_{CC} = 5$ V							
	$I_{SINK} = 8$ mA		0.1	0.25				V
	$I_{SINK} = 5$ mA					.25	.35	V
Output Voltage Drop (high)								
	$I_{SOURCE} = 200$ mA		12.5			12.5		V
	$V_{CC} = 15$ V							
	$I_{SOURCE} = 100$ mA							
	$V_{CC} = 15$ V	13.0	13.3		12.75	13.3		V
	$V_{CC} = 5$ V	3.0	3.3		2.75	3.3		V
Rise Time of Output			100			100		nsec
Fall Time of Output			100			100		nsec

NOTES:
1. Supply Current when output high typically 1mA less.
2. Tested at $V_{CC} = 5V$ and $V_{CC} = 15V$
3. This will determine the maximum value of $R_A + R_B$. For 15V operation, the max total $R = 20$ megohm.

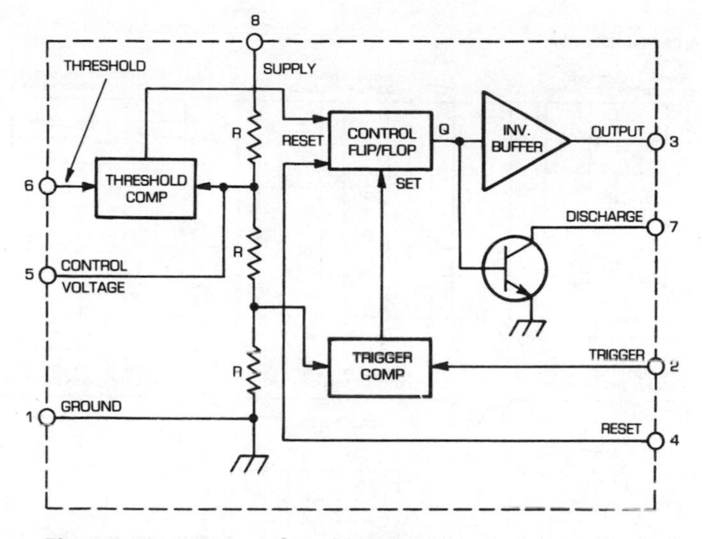

Fig. 1-7. Simplified version of the previous equivalent circuits.

voltage. Thus long-term supply variations can be ignored, and the temperature variation is only 50 ppm/°C (0.005%/°C).

In the following chapters, the circuits that will be presented will, in all cases, refer to the 8-pin 555 timer, unless otherwise specified. For those circuits using two or more timers, the 556 dual version can be used.

TYPICAL CHARACTERISTICS

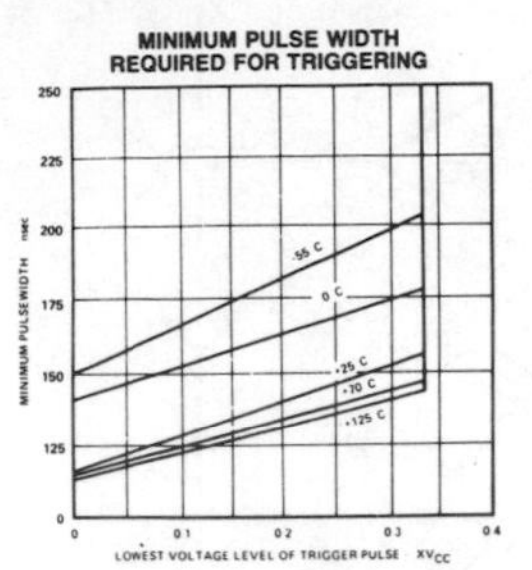

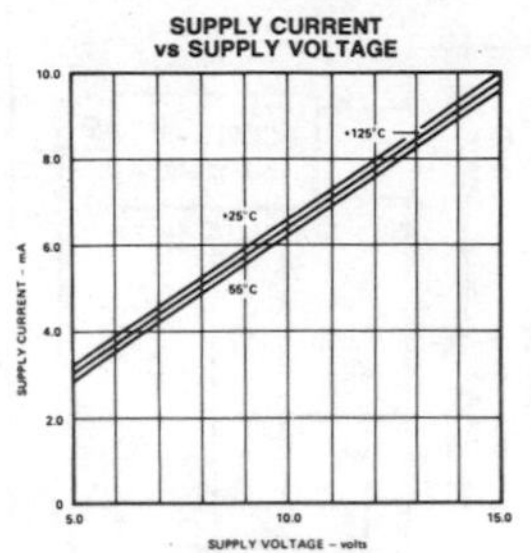

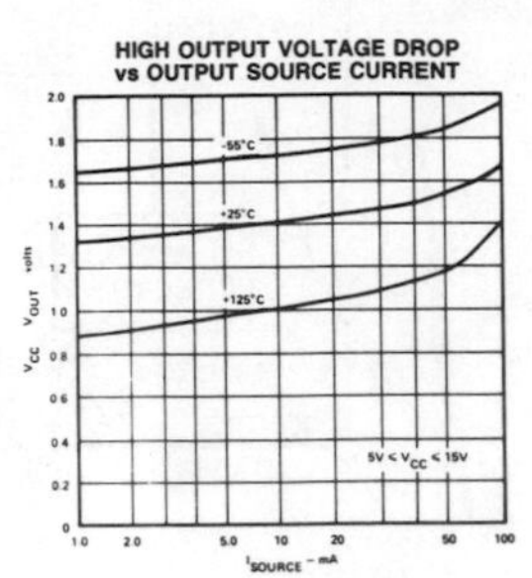

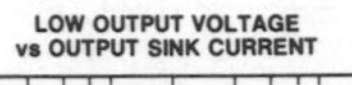

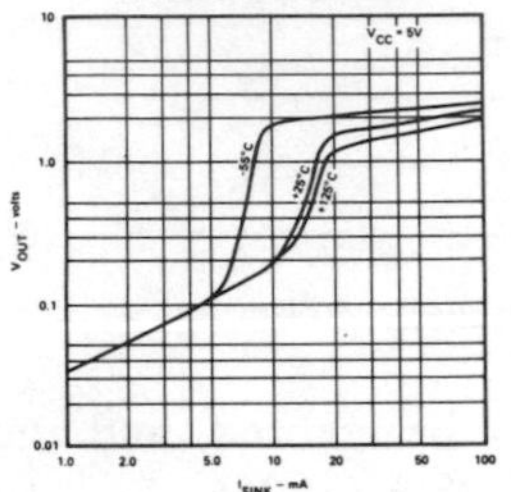

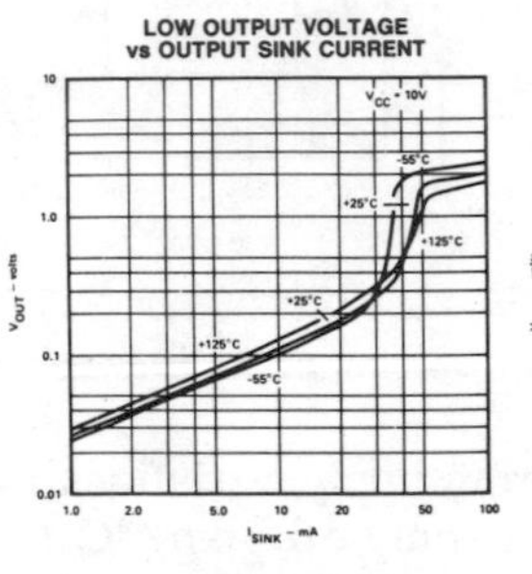

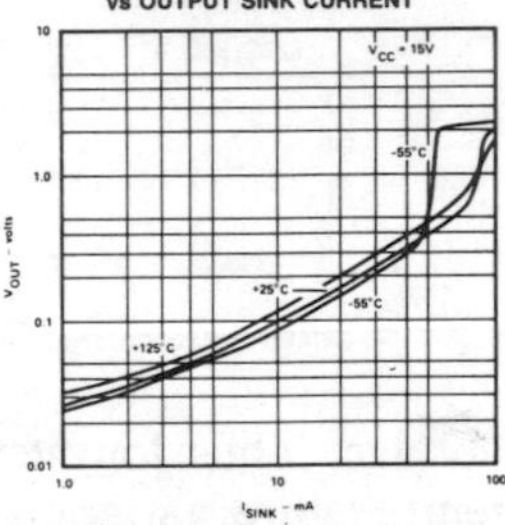

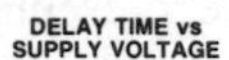

NORMALIZED DELAY TIME

SUPPLY VOLTAGE – volts

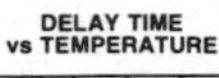

NORMALIZED DELAY TIME

TEMPERATURE – C

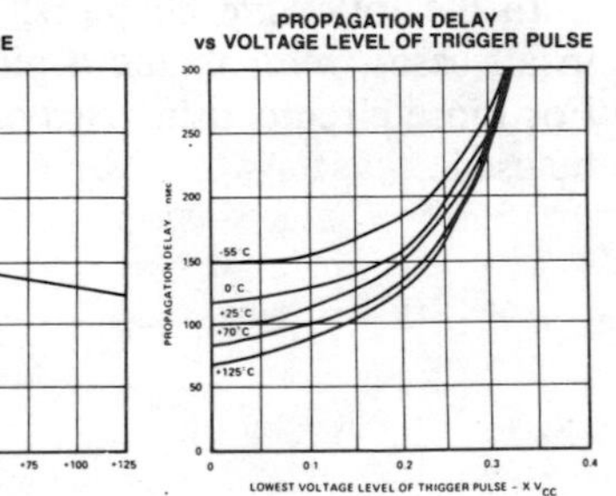

Fig. 1-8. Typical 555 timer performance curves.

CHAPTER 2

Monostable Operation

In this chapter, we will discuss the external components and connections required for the 555 timer to function as a monostable, or one-shot multivibrator.

OPERATION

The device is connected for monostable operation as shown in Fig. 2-1. In the standby state, referring to Fig. 2-2A, the control flip-flop holds Q_1 ON, thus clamping the external timing capacitor C to ground. The output (pin 3) during this time is at ground potential, or LOW. The three 5 kΩ internal resistors act as voltage dividers, providing bias voltages of ⅔ V_{cc} and ⅓ V_{cc} respectively. Since these two voltages fix the necessary comparator threshold voltages, they also aid in determining the timing interval.

Since the "lower" comparator is biased at ⅓ V_{cc}, it remains in the standby state so long as the trigger input (pin 2) is held above ⅓ the

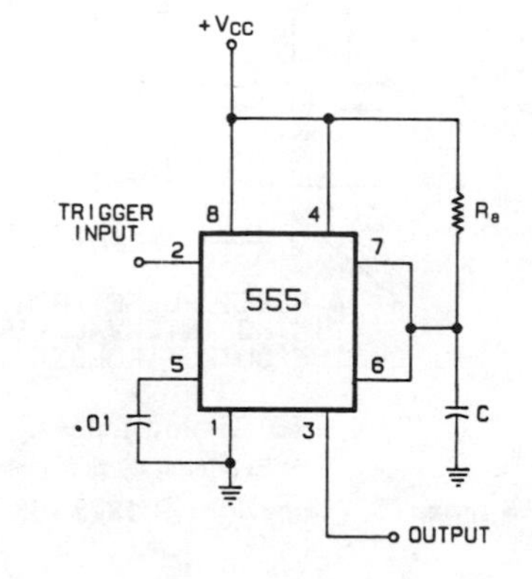

Fig. 2-1. A monostable multivibrator.

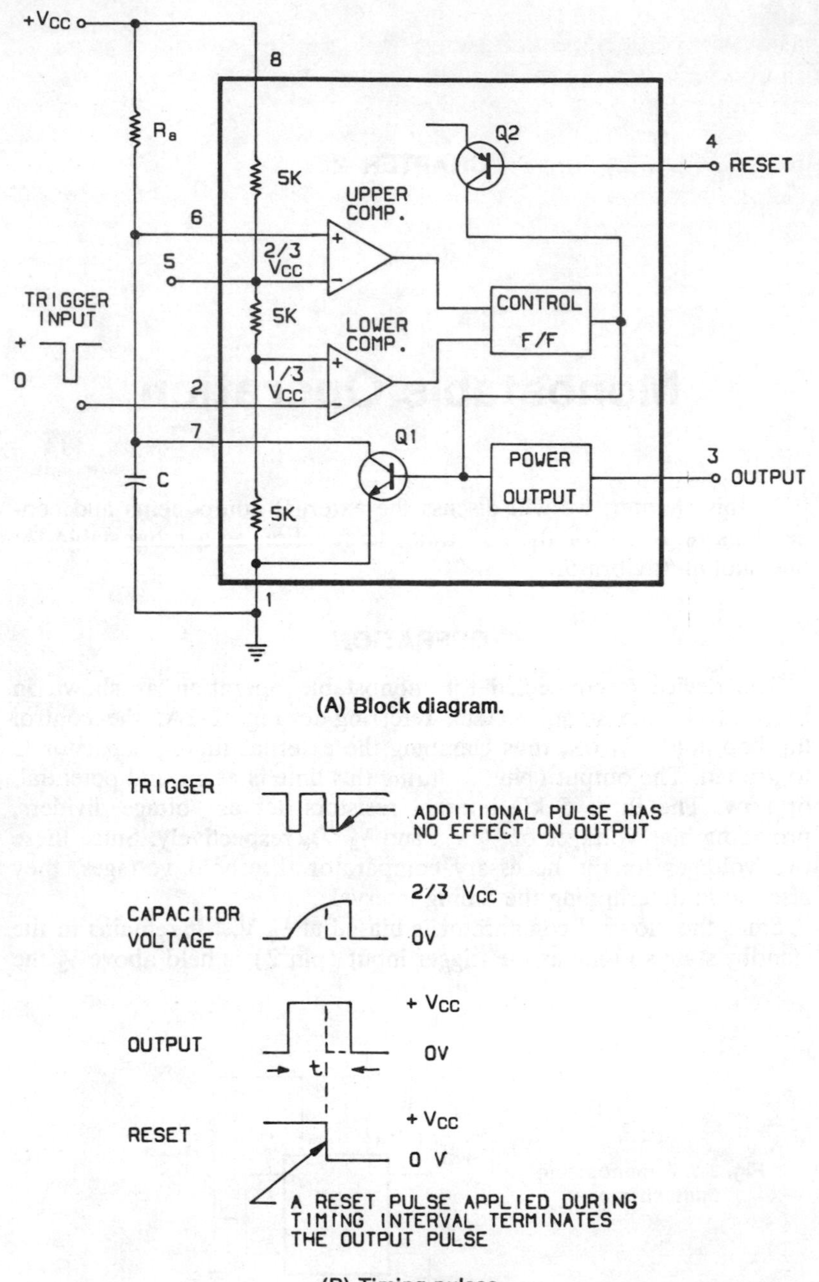

(A) Block diagram.

(B) Timing pulses.

Reprinted by permission of *Popular Electronics* magazine.

Fig. 2-2. The 555 timer IC. (Copyright © 1973 Ziff-Davis Publishing Company.)

supply voltage, V_{cc}. When triggered only by a negative-going pulse, the lower comparator sets the internal flip-flop which releases the short circuit across the timing capacitor, thus turning Q_1 OFF, and the output goes HIGH (approximately equal to V_{cc}). Since the timing capacitor is now unclamped, the voltage across it now rises exponentially through R_a towards V_{cc}, with a time constant of R_aC. After a period of time, the capacitor voltage will equal ⅔ V_{cc}, and the "upper" comparator resets the internal flip-flop, which in turn discharges the capacitor rapidly to ground potential, turning Q_1 ON. As a consequence, the output now returns to the standby state, or ground.

The 555 monostable timing sequence is shown in Fig. 2-2B. The circuit triggers only on a negative-going pulse when the level is less than ⅓ V_{cc}. Once triggered, the output will remain HIGH until the set time has elapsed, even if it is triggered again during this interval. Since the external capacitor voltage changes exponentially from 0 to ⅔ V_{cc},

$$\Delta V = V_{cc}(1 - e^{-t/R_aC})$$

$$\tfrac{2}{3} V_{cc} = V_{cc}(1 - e^{-t/R_aC}) \qquad \text{(Eq. 2-1)}$$

or

$$t = - R_aC \ln(\tfrac{1}{3})$$

so that the time width that the output is HIGH is then equal to

$$t = 1.1\ R_aC \text{ (seconds)} \qquad \text{(Eq. 2-2)}$$

Fig. 2-3 shows a graph of the various combinations of R_a and C necessary to produce a given time delay. Since the charging rate and comparator thresholds are both directly proportional to the supply voltage, the timing interval, given by Equation 2-2, is then independent of the supply voltage. Consequently, variations in the supply voltage affect both in a manner that cancels any change in the timing interval.

If, on the other hand, a negative-going pulse is simultaneously applied to the reset terminal (pin 4) and the trigger input (pin 2) during the timing cycle, the external timing capacitor is immediately discharged, and the timing cycle starts over again. The reset terminal then acts as an inhibitor. When the reset terminal is above 1 volt, the 555 timer is free to function. If the reset is taken below 0.4 V, the output is immediately forced LOW. When the reset is released, the output will still remain LOW until a negative-going trigger pulse is again applied. When the reset function is not in use, it is strongly recommended that it be connected to V_{cc} to avoid any possibility of false triggering.

Pin 5, the control voltage pin, is primarily used for filtering when the timer is used in "noisy" environments. However, by imposing a

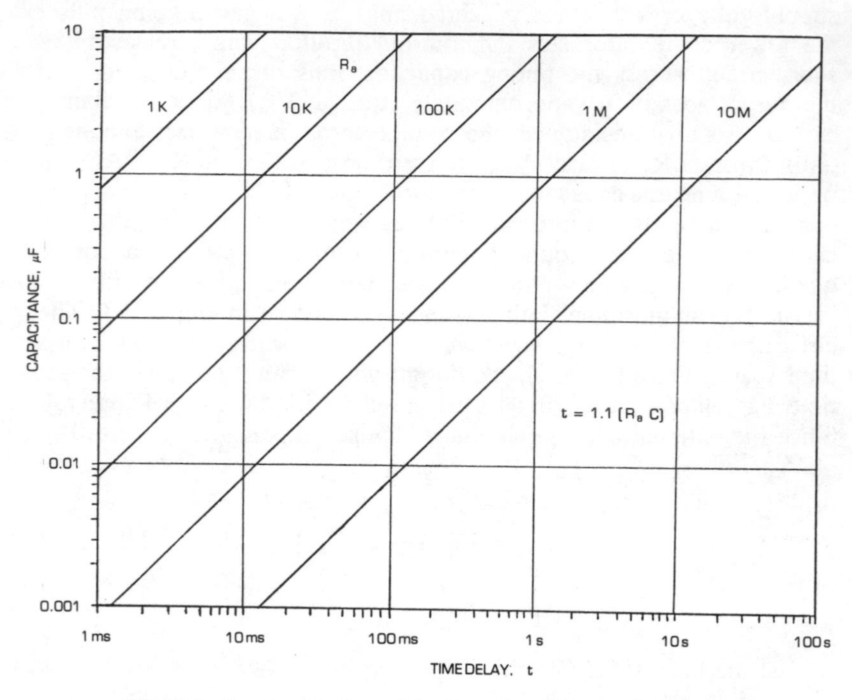

Fig. 2-3. Graph of R_aC combinations for different time delays.

voltage at this pin, it is possible to vary the timing interval, different from that given by Equation 2-2, as we shall see later in this chapter. If the control pin is not used, it is recommended that it be bypassed to ground with a 0.01 μF disc capacitor to prevent any noise from altering the calculated pulse width.

TRIGGERED MONOSTABLE

Sometimes the monostable circuit of Fig. 2-2A will mistrigger on positive pulse edges, even with the control pin bypass capacitor. To prevent any possibility of this happening, a 0.001 μF capacitor and a 10 kΩ resistor are added to the input, as shown in Fig. 2-4.

NEGATIVE-RECOVERY CIRCUIT

Ordinarily, monostable circuits require a certain amount of time to recover after triggering. If this recovery time is not completed, the next timing cycle may be shortened. The negative-recovery monostable circuit of Fig. 2-5 is often used. So long as the trigger pulse train at pin 2 keeps arriving at a certain rate, the circuit stays trig-

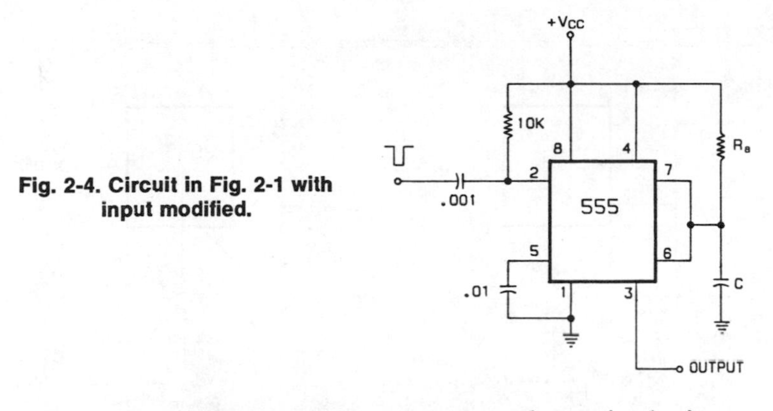

Fig. 2-4. Circuit in Fig. 2-1 with input modified.

gered, and the output remains HIGH. Any change in the input pulse frequency, or the omission of a pulse, allows completion of the timing

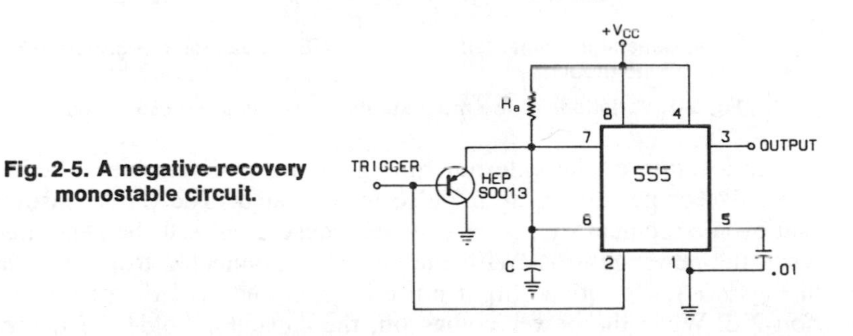

Fig. 2-5. A negative-recovery monostable circuit.

cycle and the output is forced LOW, as plotted in Fig. 2-6. As a general rule, the monostable ON time is set approximately ⅓ longer than the expected time between triggering pulses. Such a circuit is also called a missing pulse detector.

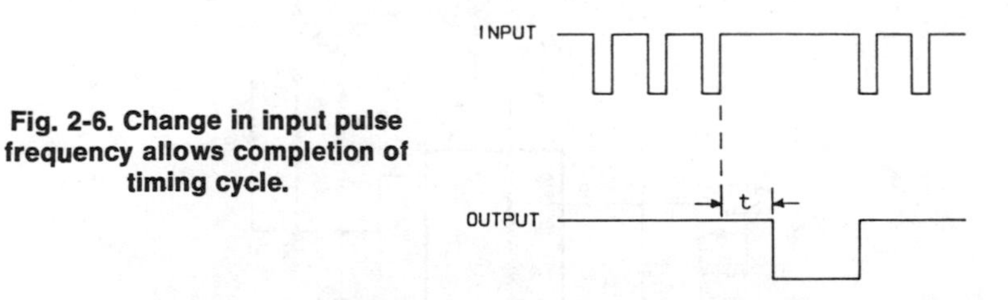

Fig. 2-6. Change in input pulse frequency allows completion of timing cycle.

Another variation of the monostable mode is to use the 555 timer to start logic circuits in their proper mode when the supply voltage is turned on, or interrupted, using the circuits shown in Fig. 2-7. An example of its use is to automatically reset TTL counters, such as the 7490, 7492, and 7493 types.

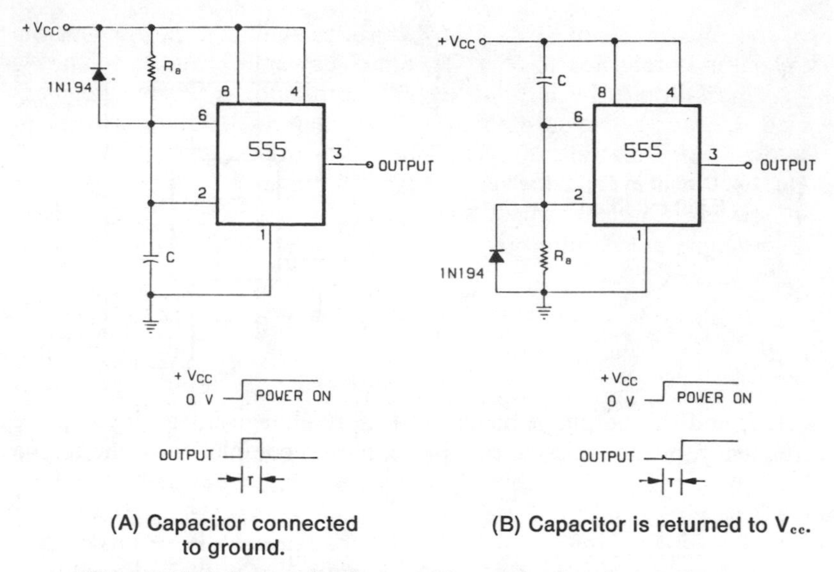

(A) Capacitor connected to ground.

(B) Capacitor is returned to V_{cc}.

Fig. 2-7. Variations of the monostable mode using the 555 timer.

The location of the external timing capacitor determines whether a positive or negative output pulse is generated. The diode ensures that even a momentary power loss will cause a pulse to be generated when the power returns. With the capacitor connected to ground, as in Fig. 2-7A, a positive output pulse is generated according to Equation 2-2. When the power comes on, the capacitor holds the trigger LOW, and the output immediately goes HIGH. When the capacitor charges to ⅔ V_{cc}, the output then goes LOW.

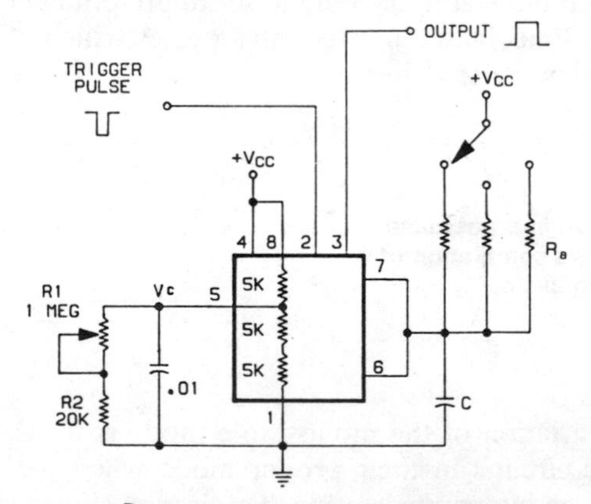

Reprinted from *Electronics*, February 6, 1975.

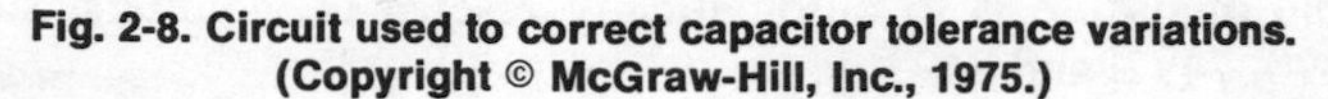
Fig. 2-8. Circuit used to correct capacitor tolerance variations. (Copyright © McGraw-Hill, Inc., 1975.)

The second circuit (Fig. 2-7B) operates similarly, except that the capacitor is returned to V_{cc}. When the power is turned ON, the ⅔ V_{cc} threshold of the internal upper comparator is immediately exceeded, forcing the output LOW. When the capacitor discharges to below ⅓ V_{cc}, the timer's output goes HIGH after a time equal to 1.1 R_aC. For this circuit, the diode across R_a assures that the capacitor will discharge quickly whenever there is a loss of power. If immediate triggering is not required, the diode can be omitted.

COMPENSATING FOR CAPACITANCE VARIATIONS

Naturally, any error in the value of the external timing capacitor causes a corresponding error in the output pulse width. If several fixed timing resistors are used to permit selection of multiple pulse widths, it may be desirable to compensate for the capacitor variation instead of "trimming" each resistor. The circuit of Fig. 2-8 allows the necessary correction for capacitor tolerance variations up to ±13% by adjustment of a single variable resistor. The output pulse width is then dependent upon the time required for the capacitor to exponentially charge to the value of the control voltage, $V_c = \frac{2}{3} V_{cc}$.

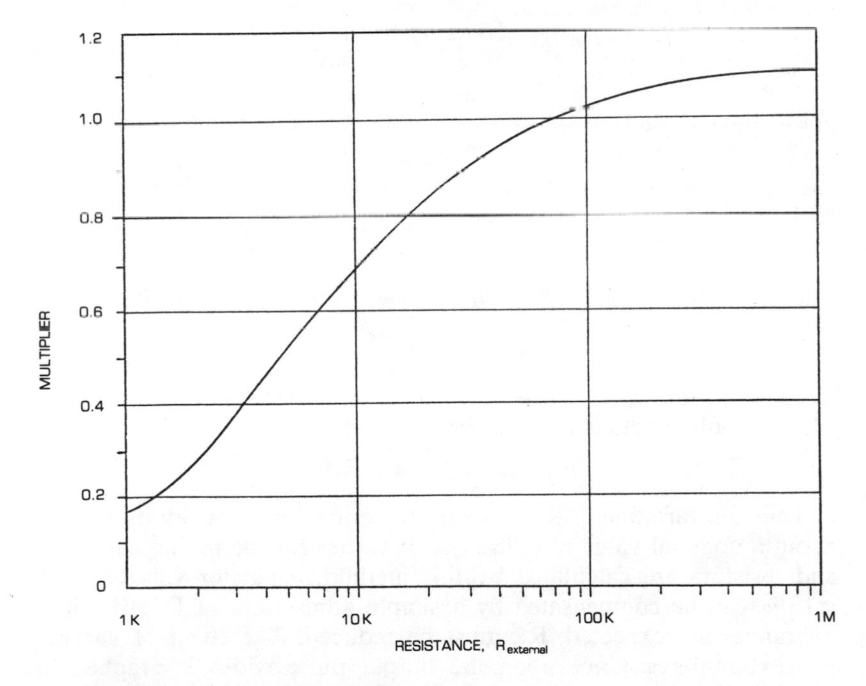

Fig. 2-9. Graph showing effect in Fig. 2-8 of R1 upon the output pulse.

This charging can then be mathematically expressed as:

$$V_c = V_{cc}(1 - e^{-t/R_aC}) \qquad \text{(Eq. 2-3)}$$

or

$$t = -R_aC \ln(1 - V_c/V_{cc}) \qquad \text{(Eq. 2-4)}$$

Equation 2-4 shows that the output pulse width is dependent on the ratio V_c/V_{cc} for any given values of the external timing components R_a and C. The technique used to compensate for capacitor variations is to vary the ratio V_c/V_{cc} with another external resistor placed in parallel with the two 5 kΩ resistors inside the timer. The external resistance is made up of a 1 MΩ potentiometer R1 in series with a 20 kΩ fixed resistor R2. The ratio V_c/V_{cc} is first determined by the voltage divider relationship,

$$\frac{V_c}{V_{cc}} = \frac{R_p}{R_p + 5\ \text{k}\Omega} \qquad \text{(Eq. 2-5)}$$

where,

R_p is the parallel combination of R1+ R2 and the 10 kΩ internal resistance, or

$$R_p = \frac{(10\ \text{k}\Omega)\ (R1 + R2)}{10\ \text{k}\Omega + R1 + R2} \qquad \text{(Eq. 2-6)}$$

When the external resistance is set at its minimum value of 20 kΩ,

$$R_p = 6.67\ \text{k}\Omega$$

and

$$V_c/V_{cc} = 0.57$$

Therefore, the output pulse width, given by Equation 2-4, is

$$t_{minimum} = 0.85\ R_aC$$

Similarly, if the external resistance is set at its maximum value of 1.02 MΩ, the pulse width is found to be

$$t_{maximum} = 1.1\ R_aC$$

Thus the variation in R1 can vary the output pulse width by ±13% about a nominal value of 0.98 R_aC. If values for the timing capacitor and resistors are calculated by this method, capacitor variations of ±13% can be compensated by a simple adjustment of R1. If wider tolerances are expected, R2 must be reduced. The effect of varying this external resistance upon the output pulse width is graphed in Fig. 2-9.

SCHMITT TRIGGERS

The 555 timer can be connected to function as a variable-threshold Schmitt trigger. Since the internal circuitry has a high input impedance and latching capability, the threshold voltage can be adjusted over a wide range with simultaneous open-collector/totem-pole outputs.

The basic internal equivalent circuits and block diagrams, given in Chapter 1, can be redrawn with logic symbols to describe the operation of the triggering circuit. As shown in Fig. 2-10, the 555 timer can be considered as a comparator that has a high input impedance that drives a Schmitt trigger, also having a high input impedance latch and a buffered strobed output.

Referring for the moment to Fig. 1-3, transistors Q_1 through Q_8 make up one of the noninverting comparators, while Q_9 through Q_{13}, and Q_{15} form the other comparator, which in turn drives the Schmitt trigger created by transistors Q_{16} and Q_{17}. Although it seems that the two comparators are simply ANDed together at the input of the Schmitt trigger, the limited source/sink current capability of the first comparator allows the other to take precedence. The first comparator then acts as a latch, allowing the other comparator, in combination with the Schmitt trigger, to be triggered when the latching, or threshold input (pin 6) is HIGH. When this input goes LOW, the Schmitt

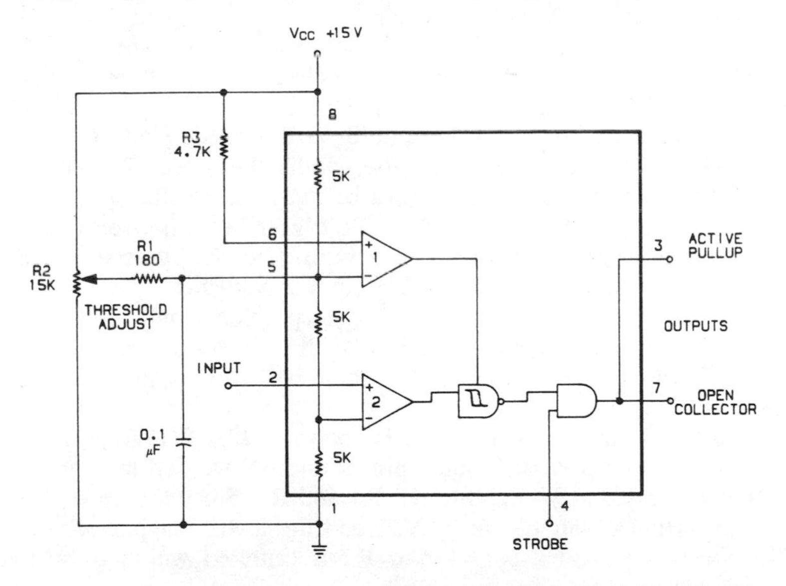

Reprinted from *Electronics*, October 25, 1973.

Fig. 2-10. Basic circuits of Chapter 1 redrawn with logic symbols. Copyright © McGraw-Hill, Inc., 1973.)

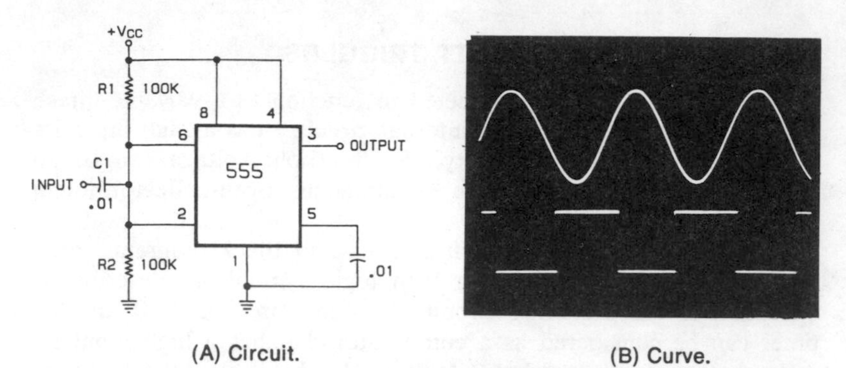

(A) Circuit. (B) Curve.

Reprinted by permission of *Popular Electronics* magazine.

Fig. 2-11. Another Schmitt trigger circuit. (Copyright © 1974 Ziff-Davis Publishing Company.)

trigger and the circuit's output are locked in whatever state the Schmitt is in.

In Fig. 2-10, resistor R3, which can range from 4 kΩ to 100 kΩ, from the latching input to V_{cc}, unlatches the Schmitt and, at the same time, tends to decouple this input from any high frequency line noise.

The 555's trigger input, which has an input impedance of about 1 MΩ, drives the Schmitt trigger, whose threshold can be varied from almost zero to just below the bias voltage existing at the latching input, by controlling the voltage at pin 5. A strobe function is provided by the timer's reset terminal; that is, the timer is active when the reset input is HIGH. Active pull-up and open collector outputs are available simultaneously at pins 3 and 7, both of which can sink up to 200 mA.

However, some precautions should be observed. Since the comparators are able to respond to pulse widths as short as 20 nsec, the control and threshold inputs should be bypassed or decoupled from the power supply. Moreover, when the trigger input is overdriven to about −0.2 V or lower, the output returns to the HIGH state, thus doubling the frequency of recurring input waveforms.

Because of the noise and bias levels, problems may arise occasionally when the control input is tied directly to the supply or to less than about 0.5 V. Resistor R1 should be 180 Ω or more to avoid these pitfalls.

Another Schmitt trigger circuit is shown in Fig. 2-11A. Here, the two internal comparator inputs (pins 2 and 6) are tied together and externally biased at $\frac{1}{2}$ V_{cc} through R1 and R2. Since the upper comparator at pin 6 will trip at $\frac{2}{3}$ V_{cc}, and the lower comparator at $\frac{1}{3}$ V_{cc}, the bias provided by R1 and R2 is centered within these two thresholds.

A sine-wave input of sufficient amplitude to exceed the reference levels causes the internal flip-flop to alternately set and reset, producing a square wave output. As long as R1 equals R2, the 555 will automatically be biased for any supply voltage in the 5 to 16 V range. From Fig. 2-11B, it should be noted that there is a 180-degree phase shift.

This circuit lends itself to condition a 60-Hz sine-wave reference signal taken from a 6.3-V ac transformer before driving a series of binary or divide-by-N counters. The major advantage of this scheme is that, unlike a conventional multivibrator type of squarer which divides the input frequency by 2, this method simply squares the 60-Hz sine-wave reference signal without division.

INVERTING BISTABLE BUFFER

Reducing the input time constant of the Schmitt circuit of Fig. 2-11, by lowering C1 to 0.001 μF, will cause the input pulses to be differentiated and the 555 timer can then be used either as a bistable device or as an inverter. In the latter case, the fast time constant, formed by C1 and R1 in parallel with R2, causes only the edges of the input pulse or square wave to be passed. These pulses set and reset the flip-flop, giving a HIGH level output that is 180 degrees out of phase.

THINGS TO REMEMBER

For proper monostable operation with the 555 timer, the negative-going trigger pulse width should be kept short compared to the desired output pulse width. Values for the external timing resistor and capacitor can either be determined from Equation 2-2 or from the graph of Fig. 2-3. However, one should stay within the ranges of resistances shown to avoid the use of large value electrolytic capacitors, since they tend to be leaky. Otherwise, tantalum or Mylar types should be used.

CHAPTER 3

Astable Operation

This chapter will present the second basic operating mode of the 555 timer as an astable multivibrator, or square wave clock.

OPERATION

The device is connected for astable operation as illustrated in Fig. 3 1. Here the timing resistor is now split into two sections, R_a and R_b, with the discharge transistor (pin 7) connected to the junction of R_a and R_b. When the power supply is connected, the timing capacitor C charges towards $\frac{2}{3}$ V_{cc} through R_a and R_b. When the capacitor voltage reaches $\frac{2}{3}$ V_{cc}, the upper comparator triggers the flip-flop and the capacitor starts to discharge towards ground through R_b. When the discharge reaches $\frac{1}{3}$ V_{cc}, the lower comparator is triggered and a new cycle is started.

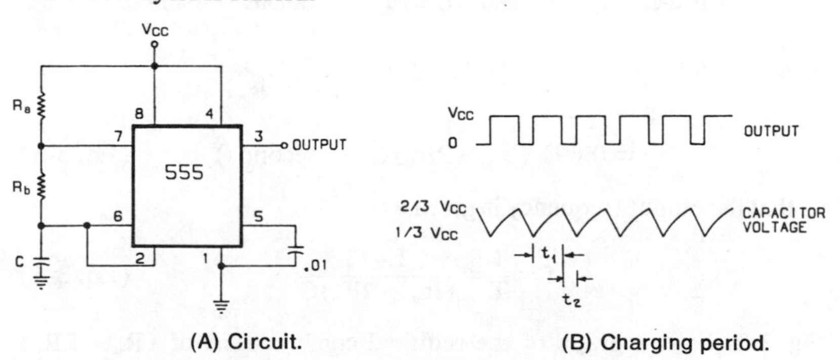

(A) Circuit. (B) Charging period.

Fig. 3-1. The 555 timer connected for astable operation.

The capacitor is then periodically charged and discharged between $\frac{2}{3}$ V_{cc} and $\frac{1}{3}$ V_{cc} respectively, as shown in Fig. 3-1B. The output state is HIGH during the charging cycle for a time period t_1, so that

$$t_1 = (R_a + R_b)C \ln\left\{\frac{V_{cc} - \frac{2}{3} V_{cc}}{V_{cc} - \frac{1}{3} V_{cc}}\right\}$$

or

$$t_1 = 0.693\ (R_a + R_b)C \qquad \text{(Eq. 3-1)}$$

The output state is LOW during the discharge cycle for a time period t_2, given by

$$t_2 = 0.693\ R_bC \qquad \text{(Eq. 3-2)}$$

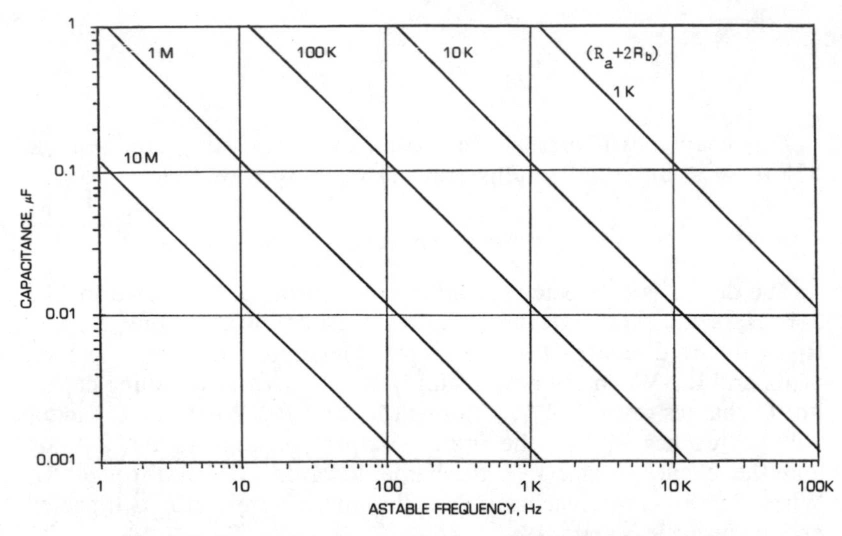

Fig. 3-2. Graph showing required combinations of R and C.

Thus, the total period charge and discharge is

$$\begin{aligned} T &= t_1 + t_2 \\ &= 0.693\ (R_a + 2R_b)C \quad \text{(seconds)} \end{aligned} \qquad \text{(Eq. 3-3)}$$

so that the output frequency is given as

$$f = \frac{1}{T} = \frac{1.443}{(R_a + 2R_b)C} \qquad \text{(Eq. 3-4)}$$

Fig. 3-2 shows a graph of the required combinations of $(R_a + 2R_b)$ and C to produce a given astable output frequency.

THE DUTY CYCLE

The duty cycle D of a recurring output is defined as the ratio of the HIGH time to the total cycle, or

$$D = \frac{t_1}{T} = \frac{R_a + R_b}{R_a + 2R_b} \quad \text{(Eq. 3-5)}$$

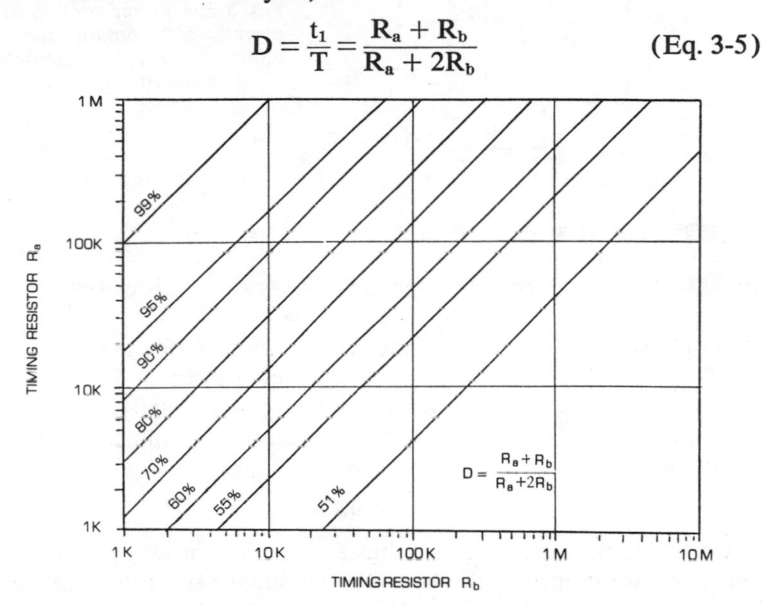

Fig. 3-3. Selection of timing resistor ratio versus duty cycle.

Thus by making R_b large with respect to R_a, we essentially can obtain a symmetrical square wave with a duty cycle of almost 50%. (See graph in Fig. 3-3.)

By using the circuit of Fig. 3-4, we can achieve a perfectly symmetrical square wave by adding a clocked flip-flop that acts as a binary divider to the timer's output. Then the selection of R_a and R_b can be

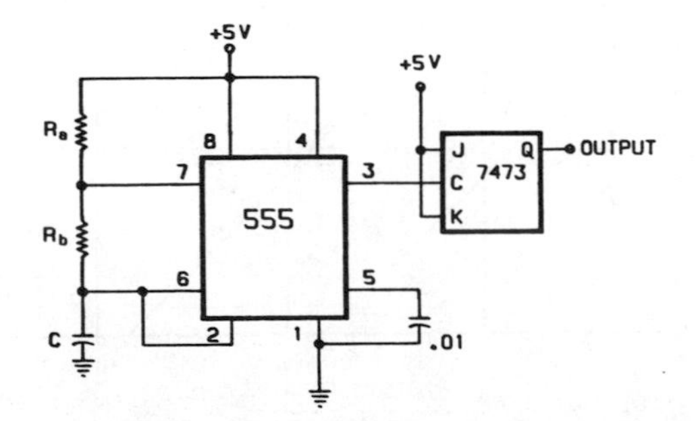

Fig. 3-4. Adding a clocked flip-flop to circuit in Fig. 3-1.

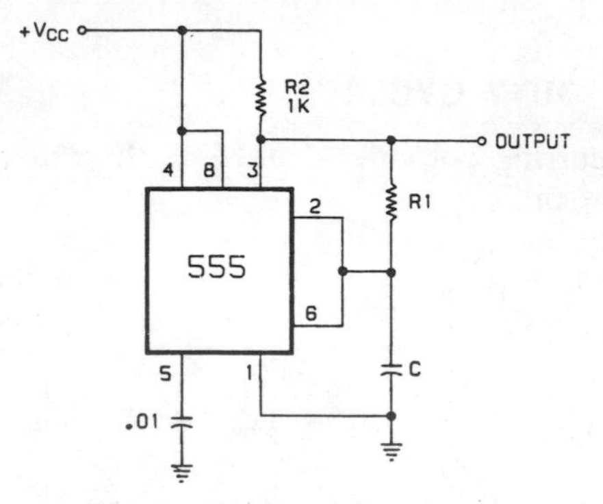

Fig. 3-5. Another version of the circuit, with timing capacitor charging to V_{cc}. (Copyright © McGraw-Hill, Inc., 1976.)

Reprinted from *Electronics,* May 13, 1976.

made without concern for the expected duty cycle. However, the output frequency of the flip-flop will now be one-half that of the timer.

On the other hand, a 50% duty cycle can be easily achieved without frequency division, by using the circuit shown in Fig. 3-5. For this circuit, the timing capacitor charges exponentially towards V_{cc} only through a single resistor R1. Consequently, the time that the output is HIGH is

$$t_1 = 0.693\,R1C \qquad \text{(Eq. 3-6)}$$

When the capacitor voltage reaches $\frac{2}{3}V_{cc}$, the output goes LOW, and the capacitor discharges through R1. The time period that the output is LOW during capacitor discharge is then

$$t_2 = 0.693\,R1C \qquad \text{(Eq. 3-7)}$$

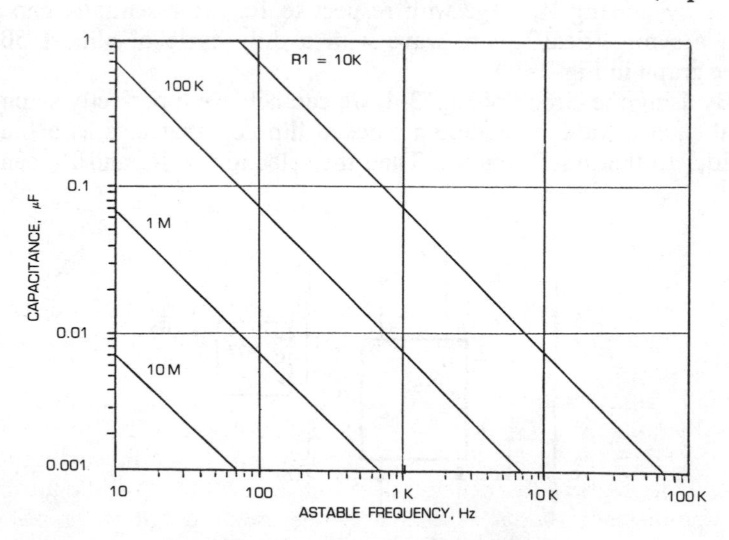

Fig. 3-6. RC timing graph for Fig. 3-5.

so that the total astable period is

$$T = t_1 + t_2$$
$$= 1.386\ R1C \qquad \text{(Eq. 3-8)}$$

and the output frequency is then

$$f = \frac{0.722}{R1C} \qquad \text{(Eq. 3-9)}$$

and is graphed in Fig. 3-6. Resistor R2 is a 1 kΩ pull-up resistor to insure that the HIGH output voltage is approximately equal to V_{cc}. R1 should be at least 10 R2, or 10 kΩ.

Sometimes, with heavy loads, the output of the basic astable circuit of Fig. 3-1A may be offset by 1 V or more from V_{cc} or ground. Since this offset potential varies the voltage across the RC timing network, the output frequency and/or the duty cycle will also be affected. The circuit of Fig. 3-7 adds a transistor and a diode to the RC timing network to permit the frequency to be varied over a wide range while maintaining a constant 50% duty cycle. When the timer's output is HIGH, Q1 is biased into saturation by R2 so that the charging current passes through Q1 and R1 to C. When the output goes LOW, the discharge transistor (pin 7) cuts OFF Q1 and discharges the capacitor through R1 and D1. Since the impedance of both paths is equal, the HIGH and LOW periods of the resultant output are also equal.

Fig. 3-7. Addition of a transistor and diode to the timing circuit.

Reprinted from *Electronics*, November 28, 1974.

Q1 should have a high beta so that R2 can be large and still cause Q1 to saturate. A high conductance germanium or Schottky diode for D1 will minimize the diode voltage drops in Q1 and D1.

For precise square waves, the ON characteristic of Q1 should be the same as that of D1 and the internal pull-down switch of the 555 timer. The method of optimizing this balance is to set the timing network to its highest frequency range and adjust R2 while monitoring the square wave output so that a symmetrical square wave will be maintained for all combinations of R1 and C.

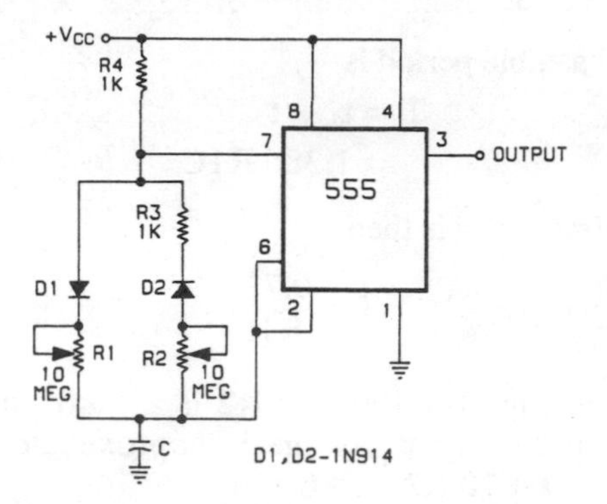

Reprinted from *Electronics*, September 19, 1974

Fig. 3-8. Circuit used to independently control the charge and discharge periods. (Copyright © McGraw-Hill, Inc., 1974.)

ADJUSTING THE FREQUENCY AND DUTY CYCLE

Several circuits are available to permit independent control over the timer's output frequency or, conversely, to make the frequency independent so that the duty cycle can be easily varied over a wide range while keeping the output pulse rate constant.

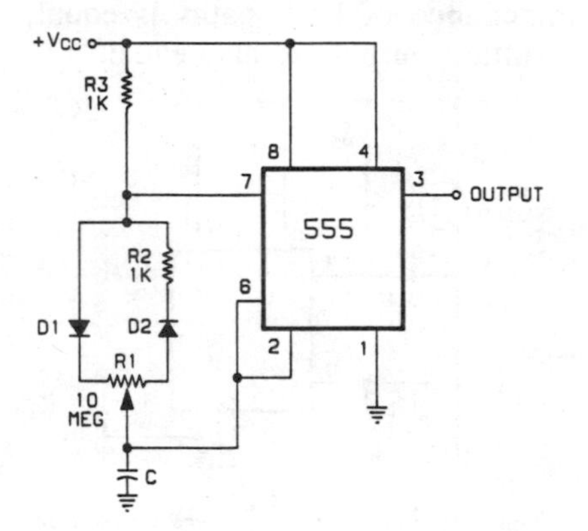

Fig. 3-9. A modification of Fig. 3-8. (Copyright © McGraw-Hill, Inc., 1974.)

Reprinted from *Electronics*, September 19, 1974.

The circuit of Fig. 3-8 is used for independent control over the charge and discharge periods. Diodes D1 and D2 provide separate paths for the timing capacitor currents. Potentiometers R1 and R2 control the HIGH and LOW periods independently over the timer's complete normal range. Resistor R3 provides the same maximum fixed resistance in the charge loop as R4 gives in the discharge loop.

The circuit of Fig. 3-9, which is a modification of Fig. 3-8, makes the period dependent. As R1 is varied, one period is decreased while the other is increased proportionately. For example, if R1 is 10 MΩ, and R2 and R3 are both 1 kΩ, the duty cycle will range from about 0.01% to 99.99% with little change in frequency.

In both circuits, the voltage drop across the diodes decreases the effective voltage across the RC timing network. Consequently, the output periods will be less than those given by Equations 3-1 and 3-2. For these circuits, the constant voltage drops of the diodes must be accounted for. If the voltage drop across each diode is 0.6 V in the forward direction, then the timer's period when the output is HIGH is

$$t_1 = RC\ ln\left\{\frac{V_{cc} - \frac{1}{3}V_{cc} - 0.6}{V_{cc} - \frac{2}{3}V_{cc} - 0.6}\right\}$$

or

$$t_1 = RC\ ln\left\{\frac{\frac{2}{3}\ V_{cc} - 0.6}{\frac{1}{3}\ V_{cc} - 0.6}\right\} \qquad \text{(Eq. 3-10)}$$

Therefore, the lower the supply voltage, the greater the effect of the diode's voltage drop.

A constant period with variable duty-cycle astable timer can be obtained with the circuit of Fig. 3-10, using transistors instead of diodes. Q1 and Q2 are both in saturation when the output of the timer is HIGH. During this time C is being charged by the circuit through Q1, R1, and R_a. When the voltage across C reaches $\frac{2}{3}\ V_{cc}$, the output goes LOW. The capacitor now discharges through R2 and R_b until the lower threshold of $\frac{1}{3}\ V_{cc}$ is reached, and the output goes HIGH. Then the cycle starts over again.

The time when the output is HIGH, during the charging cycle, is

$$t_1 = 0.693(R1 + R_a)C \qquad \text{(Eq. 3-11)}$$

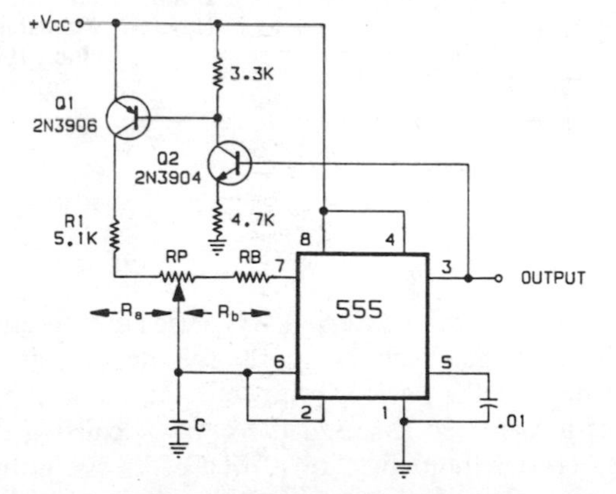

Reprinted from *Electronic Design*, July 5, 1975.

Fig. 3-10. Variable duty-cycle astable timer circuit. (Copyright © Hayden Publishing Company, Inc., 1975.)

and the output is LOW for

$$t_2 = 0.693(R2 + R_b)C \qquad \text{(Eq. 3-12)}$$

so that the total period is

$$T = 0.693(R1 + R2 + R_a + R_b)C \qquad \text{(Eq. 3-13)}$$

Since the quantity $(R_a + R_b)$ equals the potentiometer resistance RP, then

$$T = 0.693(R1 + R2 + RP)C \qquad \text{(Eq. 3-14)}$$

Consequently, the duty cycle is

$$\frac{R1 + R_a}{R1 + R2 + RP} \qquad \text{(Eq. 3-15)}$$

The component values shown in Fig. 3-10 allow the duty cycle to be adjusted between 1% and 99%.

A CRYSTAL-CONTROLLED ASTABLE

Although the basic astable circuit of Fig. 3-1A is accurate for most applications, the circuit of Fig. 3-11 enables the 555 timer to be changed from an RC to a stable crystal-controlled oscillator. The

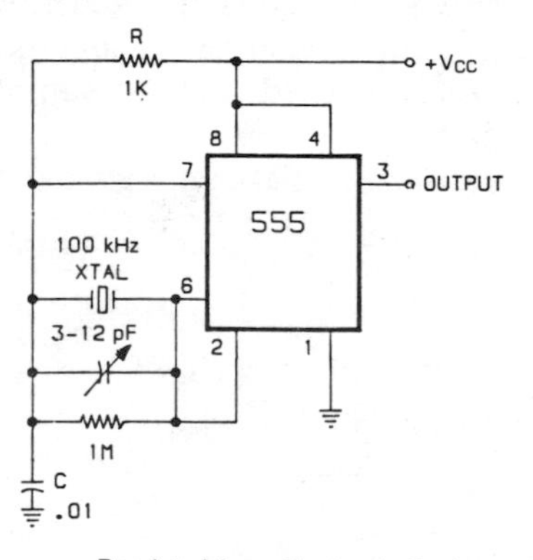

Reprinted from *Electronic Design*, November 8, 1974.

Fig. 3-11. A stable crystal-controlled oscillator. (Copyright © Hayden Publishing Company, Inc., 1974.)

crystal is placed between the external RC series circuit and the timer's comparator. The charge/discharge paths for the capacitor remain the same as described for the basic circuit, but the control signal to both comparators is now forced through the crystal, causing the circuit to oscillate at the crystal frequency, or at one of its subharmonics.

The values for R and C are selected so that, with the crystal shorted out, oscillation will still be in the vicinity of the crystal frequency, or

$$f = \frac{1.443}{RC} \qquad \text{(Eq. 3-16)}$$

But the values for R and C can vary by 25% or more without the crystal's frequency being affected, since the charge/discharge amplitudes of the capacitor voltage change to accommodate the selected values for R and C, thus keeping the frequency constant.

If we double the time constant while using the same crystal, the oscillations shift to one-half the crystal frequency. Other changes also produce corresponding subharmonics—$\frac{1}{3}$, $\frac{1}{4}$, $\frac{1}{5}$, etc.—of the crystal frequency.

The variable capacitor across the crystal allows precise control of the crystal frequency against a known standard, such as the broadcasts by WWV. The 1 MΩ resistor provides a dc path for the comparator inputs to ensure that the oscillator will start when the power is first applied.

MINIMIZING DIFFERENCES IN THE MONOSTABLE AND ASTABLE PERIODS

A simple component, such as a diode or a resistor, can minimize the difference between the monostable and astable timing periods of the 555 timer. When the timer is used as an astable multivibrator, the output, as given by Equation 3-3, has a period equal to 0.693 RC. However, when wired as a monostable multivibrator, or when strobed via the reset input, a longer period of 1.1 RC (Equation 2-2) results.

In the circuit of Fig. 3-12, the conventional circuit arrangement for the timer is shown. The switch selects either the monostable or astable mode. Either a resistor or a diode may be added to equalize the timing periods. The diode, placed between pins 3 and 5, pulls down the

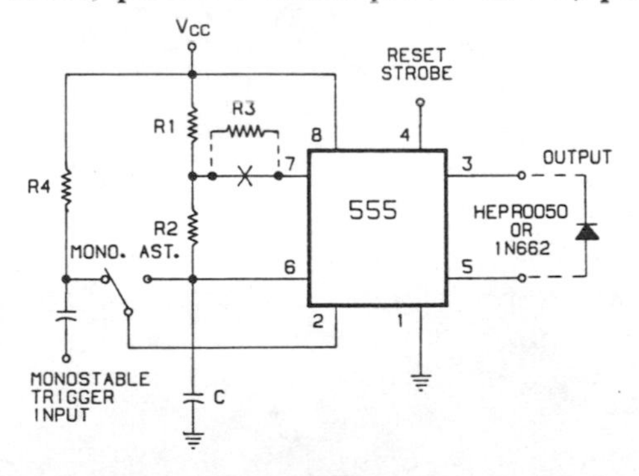

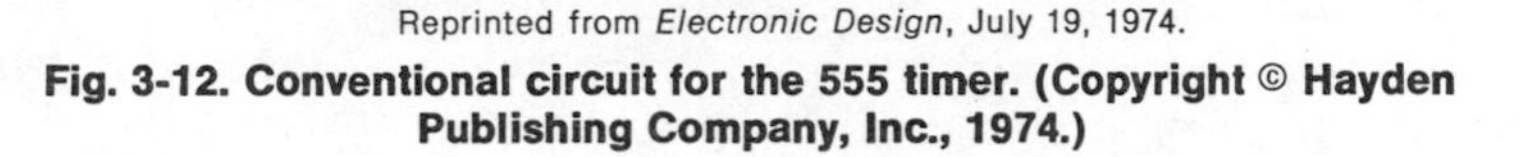
Reprinted from *Electronic Design*, July 19, 1974.

Fig. 3-12. Conventional circuit for the 555 timer. (Copyright © Hayden Publishing Company, Inc., 1974.)

control voltage to about 0.9 V each time the output goes LOW. Therefore, the timing capacitor C must drop to about 0.5 V before the level at pin 2 can trigger another output pulse. The two periods will agree to within 5% since the capacitor starts to charge from near ground level in both modes.

In the second method, resistor R3 forces the monostable period to approach that of the astable when it prevents the capacitor from discharging completely. Adjustment of the voltage divider (formed by R1 and R3) permits the capacitor voltage to drop only far enough to trigger a new pulse.

The advantage of the diode method is that no calculations are required to provide close matching of the pulse widths. Also, a single potentiometer control of the pulse width is still possible. However, the lower threshold, and consequently the pulse width, depends on the diode's offset and drift characteristics.

The advantage of the resistor method is that the periods of the two modes are controlled by R1 and R3. Thus, the periods can be set very close to each other, and a 0.01 μF bypass capacitor to ground can be placed at pin 5. Also the resistor method does not introduce the temperature variations of a diode, and the matching of the pulse widths tends to remain constant with variations in V_{cc}. However, one disadvantage is that R1 cannot be varied to control the pulse period without also adjusting R3.

With this chapter, we have concluded the discussion covering the basic modes of operation of the 555 timer. In the following chapters, we will realize how versatile this device is in fulfilling the requirements of a myriad of applications.

CHAPTER 4

Power Supply Circuits

Many times, the dc voltage needed to operate electronic equipment is different from the dc supply that is available. A converter circuit is used to convert direct current flow from one level to another.

POSITIVE DC-DC CONVERTERS

The transformerless dc-dc converter shown in Fig. 4-1 may be adapted to a variety of low-power applications requiring higher dc voltages than the available supply. The 555 timer is connected as a free-running square wave generator at approximately 3 kHz, and a capacitor-diode voltage-doubler section is added to the output. The input 0.01 μF capacitor is used to help filter out the resultant 3 kHz signal from entering the input power lines. Additional doubler sections, as shown in Fig. 4-2, may be added to increase the voltage output, but only at the expense of available current. The input voltage can be in the range of 4.5 V to 16 V.

An improved dc-dc converter with a current step-up regulator and choke input filtering is shown in Fig. 4-3. As before, the timer operates in the astable mode, but transistors Q2 and Q3 are added to keep the output filter capacitor charged to the desired output voltage. In addition, any overvoltage is prevented by feedback that turns off the timer when the filter capacitor's voltage reaches a predetermined level.

Since the maximum operating voltage will be 16 V for most 555-type timers, V_{cc} is clamped at the voltage of the zener diode DZ1, so, that the regulated output voltage is approximately

$$V_{out} = V_{Z2} + 0.3 \qquad \text{(Eq. 4-1)}$$

where,

V_{Z2} is the zener voltage of DZ2.

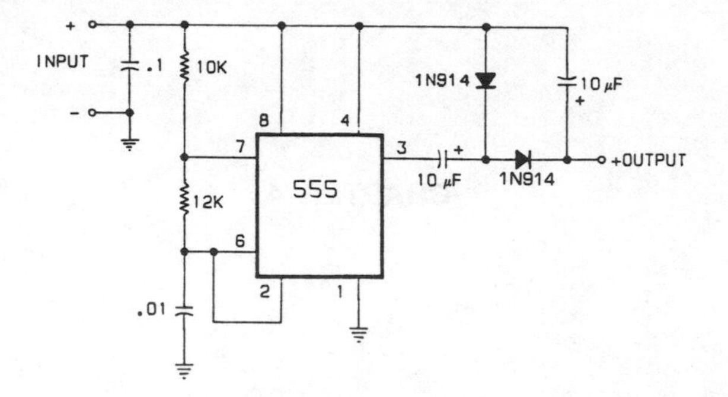

Reprinted by permission of *Ham Radio* magazine.

Fig. 4-1. A transformerless dc-dc converter. (Copyright © 1975 Communications Technology, Inc.)

When the timer's output is HIGH, both Q2 and Q3 are turned ON so that the collector current from Q3 flows through the 1 mH choke to the load and the filter capacitor. When the timer's output is LOW, Q2 and Q3 are OFF, and D1 commutates the current flowing through the choke.

The feedback circuit consists of R4, DZ2, Q1, and R3. Whenever the output voltage exceeds ($V_{Z2} + 0.3$), Q1 turns ON and drives the timer's reset input LOW, causing Q2 and Q3 to remain OFF, allowing the output voltage to decrease. If there were no feedback circuit, the output voltage would depend entirely upon the input voltage and duty cycle, so that

$$V_{out} = V_{in}D = V_{in}\frac{t_1}{t_1 + t_2} \qquad \text{(Eq. 4-2)}$$

With the component values shown in Fig. 4-3, the performance of the circuit would be expected to give the characteristics listed in Table 4-1.

By using the modulating capability of the 555 timer, the device can be pulse-width modulated to provide current foldback in a regu-

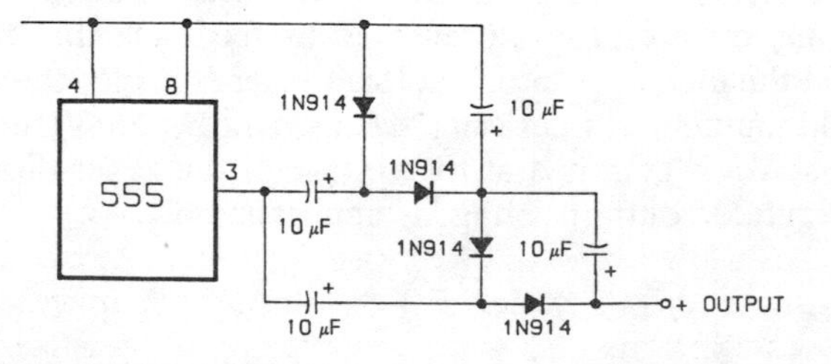

Fig. 4-2. Additional doubler sections are added to circuit of Fig. 4-1.

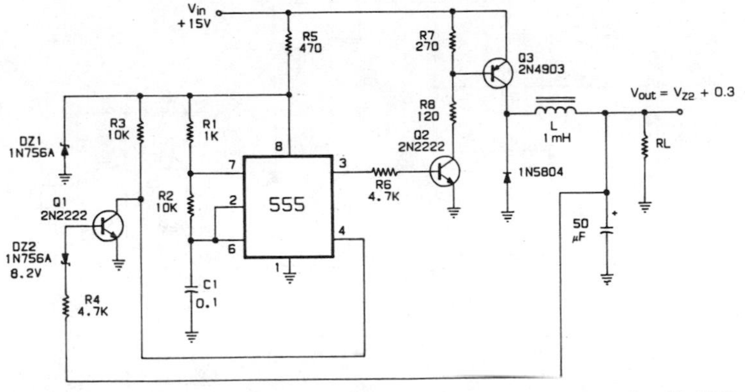

Fig. 4-3. An improved dc-dc converter. (Copyright © McGraw-Hill, Inc., 1975.)

Table 4-1. Circuit Characteristics for Fig. 4-3

Parameter	Symbol	Test Conditions	Value	Units
Input voltage	V_{in}		15	V
Output voltage	V_{out}		8.4	V
Load current	I_{out}		300	mA
Ripple current	I_r	$I_{out} = 300$ mA	5	mA
Load regulation		$V_{in} = 15$ V, $I_{out} = 0$–300 mA	0.5	%
Line regulation		$V_{in} = 15$–25 V, $I_{out} = 300$ mA	2.5	%

lated supply, as seen in Fig. 4-4. When there is no external voltage at the modulating or control input (pin 5), the device will trigger itself, producing a square wave, then amplified by Q1. Q2 will be ON as long as the timer's output is HIGH, thus driving current into R2 and C2. When Q2 is OFF, the energy stored in L and C2 is now available to supply the load.

When V_{out} differs from V_{cc}, the generated voltage is fed to Q5, acting as a comparator controlled by zener diode DZ. Q5 will not conduct unless the output voltage at the collector of Q2 continuously changes, depending on how it compares with the zener voltage. Since Q5's collector voltage feeds the modulating input of the 555, the pulse width of the generated square wave is modulated to provide the necessary output, as given by Equation 4-2.

Resistor R7 functions as a current sensor. When the load current increases to a level such that the voltage drop across R7 turns Q3 ON, Q4 will be driven to saturation, causing the timer's reset pin to be LOW, resetting the 555 with zero output. When this occurs, Q4 is OFF, and the timer is activated with the output HIGH.

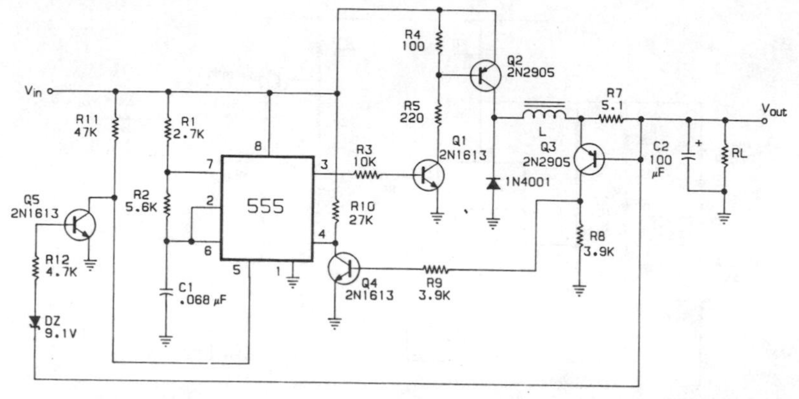

Reprinted from *EDN,* January 5, 1976.

Fig. 4-4. Circuit designed to use modulating capability of the 555 timer. (Copyright © Cahners Publishing Company, Inc., 1976.)

If an overload condition still exists, both Q3 and Q4 will again be switched ON and the timer resets. Thus, a close-loop chain reaction takes place. With a 15-V input, the circuit will deliver a 10-V, 100-mA output with line and load regulation of 0.5% and 1% respectively.

NEGATIVE DC-DC CONVERTERS

By essentially reversing the polarity of the diodes and capacitors of the circuit of Fig. 4-1, we obtain a negative output with respect to ground, as shown in Fig. 4-5. However, if a transistor is included to vary the control voltage at pin 5, a constant output voltage can be assured.

In Fig. 4-6, transistor Q1 varies the timer's control voltage, thus increasing or decreasing the timer's pulse frequency. Resistor R3 acts as a collector load for Q1 whose base is driven by R4, adjusted to compare the output voltage to the supply voltage. If the output becomes less negative, the control voltage drops closer to ground caus-

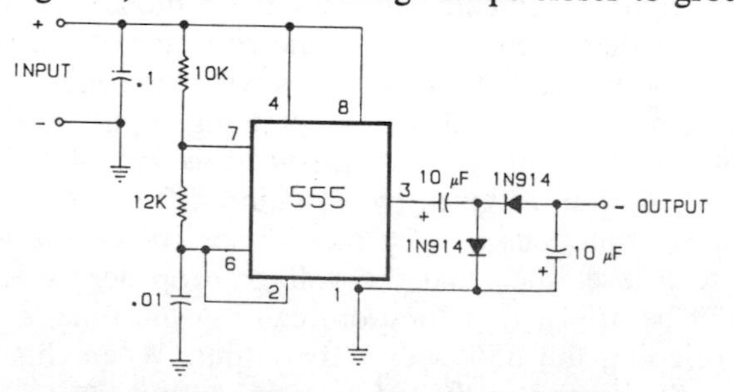

Fig. 4-5. A converter with negative output.

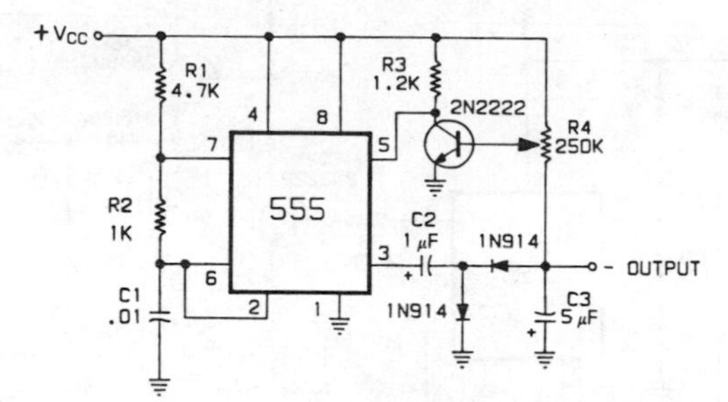

Reprinted from *Electronics*, May 13, 1975.

Fig. 4-6. Transistor added to circuit to vary the control voltage. (Copyright © McGraw-Hill, Inc., 1975.)

ing the pulse rate to increase, and C3 is charged at a faster rate. If, on the other hand, the output becomes more negative, the control voltage approaches V_{cc} and the pulse rate decreases.

Another improvement upon the basic circuit is the feedback circuit of Fig. 4-7 which will develop a −15-V output that is regulated to within 1% for no-load currents up to 30 mA. When the timer's output goes positive, C1 is charged through D1, with D2 being reverse biased. When the output goes negative, some of the charge on C1 is transferred to C2 through D2, since D1 is now reverse biased.

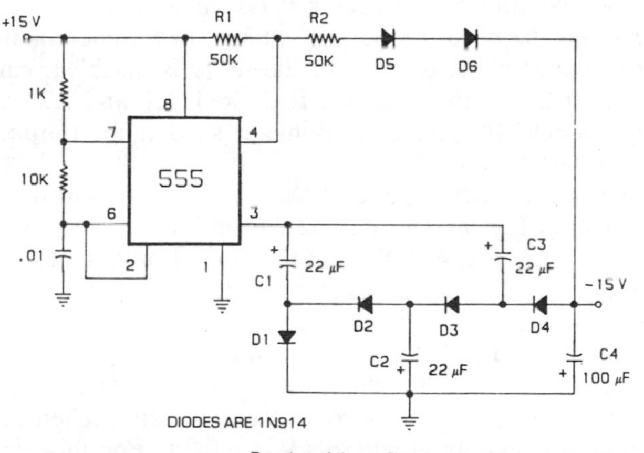

Reprinted from *Electronics*, August 22, 1974.

Fig. 4-7. Feedback circuit added to the basic circuit. (Copyright © McGraw-Hill, Inc., 1974.)

As the output again goes positive, C3 charges through C2 and D3 so that the output is now approximately twice the supply voltage, or $2V_{cc}$. When the timer's output again goes negative, this charge is transferred to C4 by D4, thus doubling the output voltage. Conse-

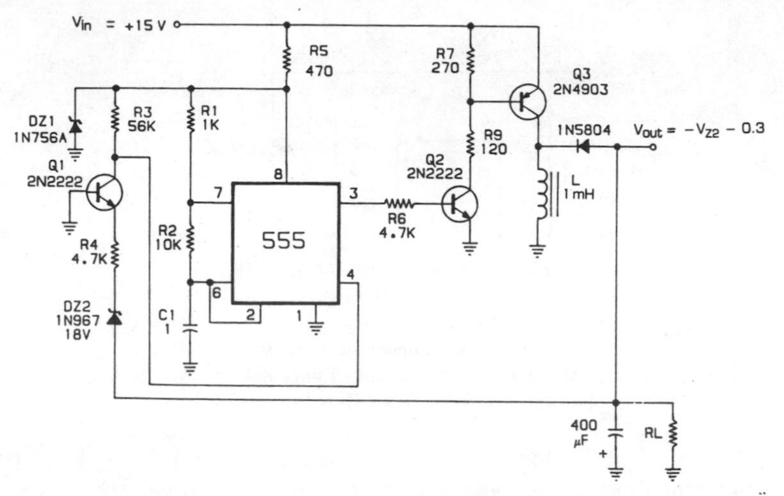

Reprinted from *Electronics*, November 13, 1975.

Fig. 4-8. Fig. 4-3 changed to a negative-polarity supply circuit. (Copyright © McGraw-Hill, Inc., 1975.)

quently, this arrangement requires the supply to both source and sink current.

Without the feedback connection, consisting of R1, R2, D5, and D6, the output with no load will be about −29 V. However, this resistor-diode combination places a 0.7-V level at the reset input of the timer when the negative output voltage magnitude equals V_{cc}. If the output voltage exceeds −15 V, the timer is disabled, causing no signal to reach the doubler circuit. If desired, R1 and R2 can be replaced by a single 100 kΩ potentiometer so that the output voltage can be made variable.

By changing the feedback and choke input elements of the positive-polarity circuit of Fig. 4-3, we can transform it into a negative-polarity supply, as shown in Fig. 4-8. When Q3 switches OFF, the commutating current in the 1 mH choke charges capacitor C to produce an output voltage that is negative with respect to ground. This voltage is then applied to Q1 through DZ2 and R4. Whenever the output is more negative than $-(V_{Z2} + 0.3)$, the timer's reset goes LOW, allowing the voltage across capacitor C to become more negative. Then the output voltage is approximately equal to $(V_{Z2} + 0.3)$. For this circuit, the negative voltage can be either equal to, less than, or greater than the supply voltage, in magnitude.

A DUAL-POLARITY CONVERTER

For a number of applications, TTL digital systems sometimes require the use of operational amplifiers which normally use a ±15-V supply, although TTL devices require only +5 V. The dual-polarity dc-dc converter, shown in Fig. 4-9, can supply the necessary ±15 V at 10 mA from an existing +5-V TTL supply.

The 555 astable frequency, with the components shown, is about 100 kHz with a duty cycle of 75%. The output drives Q1, switching the current through the primary of the pulse transformer (Pulse Engineering PE-3843 or equivalent) ON and OFF. When the current is switched OFF, a spike of about 20 V appears at the collector of Q1, which is then rectified, filtered, and regulated to give +15 V.

Simultaneously, a voltage spike is present across the transformer's secondary. Because the transformer provides dc isolation, the higher-voltage end of the transformer can be safely grounded to produce a negative pulse which is also rectified, filtered, and regulated to yield −15 V.

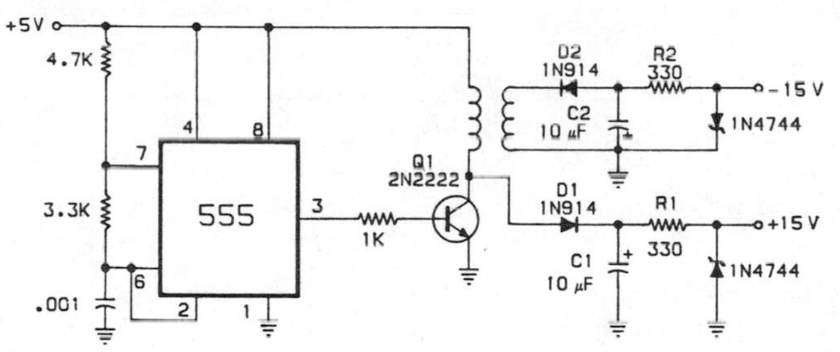

Reprinted from *Electronics,* June 12, 1975.

Fig. 4-9. A dual-polarity dc-dc converter. (Copyright © McGraw-Hill, Inc., 1975.)

BATTERY CHARGER/MONITOR

The 555 timer can also function as the heart of an automatic battery charger by using the circuit of Fig. 4-10. This circuit is intended

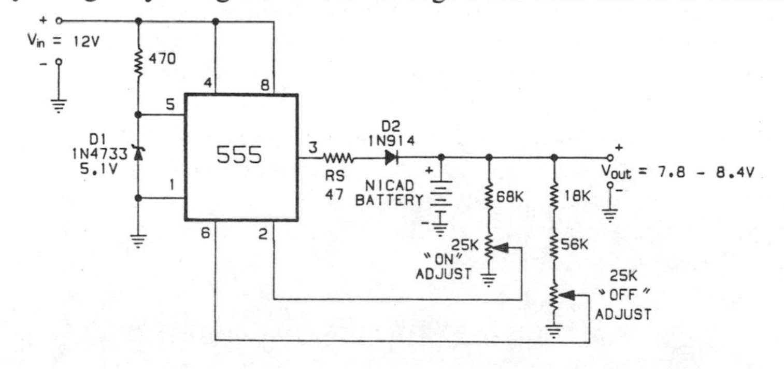

Reprinted from *Electronics,* June 21, 1973.

Fig. 4-10. A battery charger circuit. (Copyright © McGraw-Hill, Inc., 1973.)

to maintain a full charge on a standby battery supply for an instrument that is always connected to the ac power line, whether in use or not.

The necessary reference voltage for both comparators is provided by zener diode D1 through the timer's internal resistance network. Therefore, the output of the timer switches between ground and +10 V.

The circuit is calibrated by substituting a variable dc power supply for the NiCad batteries. The "off" adjustment potentiometer is set for the desired battery cutoff voltage, typically 1.4 V per cell; and the "on" adjustment pot is set for the desired turn-on voltage, about 1.3 V per cell. Resistor RS limits the circuit's operating current to less than 200 mA under all conditions. Diode D2 prevents the battery from discharging through the timer when it is turned OFF.

CHAPTER 5

Measurements and Control

This chapter is devoted to circuits using the 555 timer for the testing of other 555 devices, logic circuits, multiwire cables, the measurement of inductance, capacitance, and resistance, temperature control, tachometers, oscilloscope control and display, and waveform generators.

GO/NO GO TESTER

Now that a number of 555 timer circuits have been discussed, it may have occurred to some of you how it would be possible to rapidly test any of the devices without the use of an oscilloscope or other costly instrument. The Go/No Go circuit of Fig. 5-1 will do just that, especially when large volume testing is required.

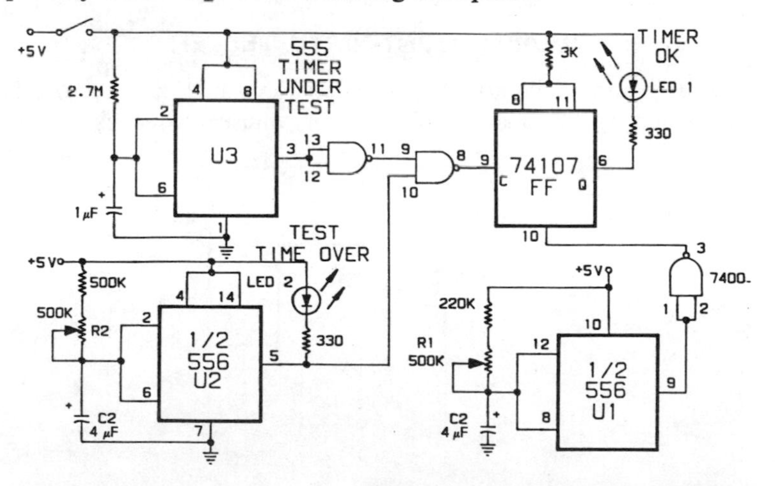

Reprinted from *Electronic Design*, May 24, 1974.

Fig. 5-1. Go/No Go test circuit. (Copyright © Hayden Publishing Company, Inc., 1974.)

Using the 556 dual timer, both sections and the timer under test are switched to the HIGH state to begin their cycles when power is first applied. The output of U1, connected as a monostable one-shot, inhibits the flip-flop for the first timing interval of 2 seconds (Fig. 5-2). This is set by R1 and C1 according to Equation 2-2. Three seconds after the power is applied, the output of U2 goes LOW and thus inhibits any signal from the timer under test (U3).

The period between 2 and 3 seconds (in Fig. 5-2) is the time allotted for U3 to complete its cycle, as indicated by a LOW output which turns LED 2 ON. Only during this time can the HIGH-to-LOW transition of U3 trigger the flip-flop, so that LED 1 is ON. LED 2 is always ON, signifying that the test is completed.

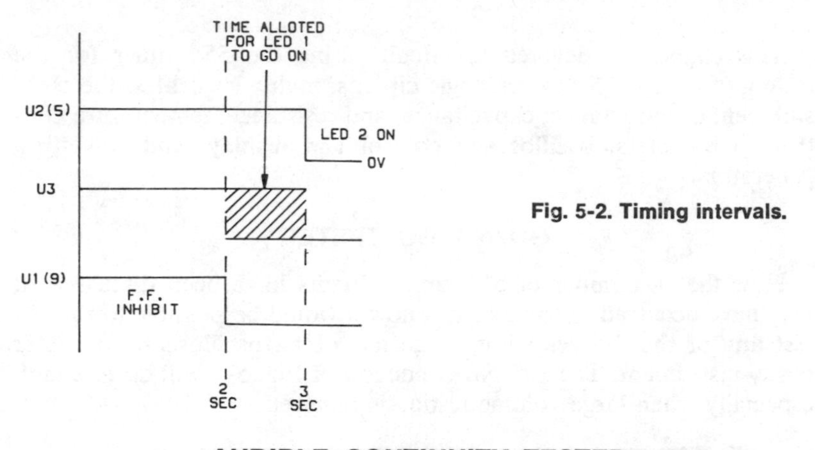

Fig. 5-2. Timing intervals.

AUDIBLE CONTINUITY TESTER

A simple audible continuity tester is shown in Fig. 5-3. The 555 is connected as an astable multivibrator whose frequency is set to

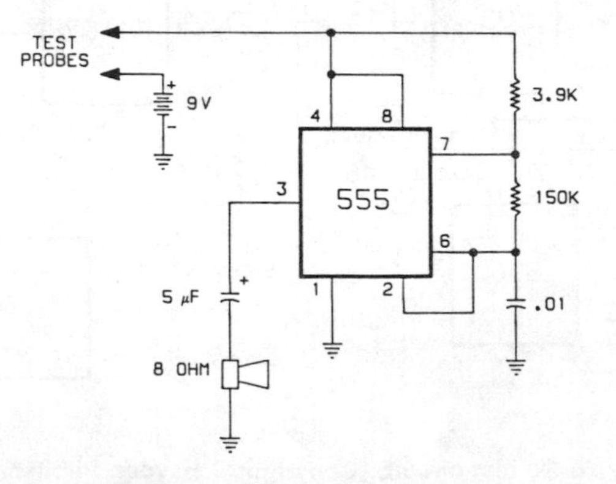

Fig. 5-3. Audible continuity tester.

operate in the 500 Hz to 2 kHz range using Equation 3-4. An inexpensive permanent-magnet 8Ω speaker, salvaged from an old transistor radio, is capacitively coupled to the timer's output.

DIGITAL LOGIC PROBES

One of the most frustrating problems facing the experimenter is the capability to check the states of TTL or CMOS devices without having a triggered scope. An inexpensive logic probe using the 555 timer for detecting and indicating logic states is shown in the schematic diagram of Fig. 5-4.

The input logic signal triggers the timer, using two diodes to keep the input signals from driving the timer's trigger input to either V_{cc} or ground. The timer then acts as a comparator, and the output assumes an inverted state relative to the input.

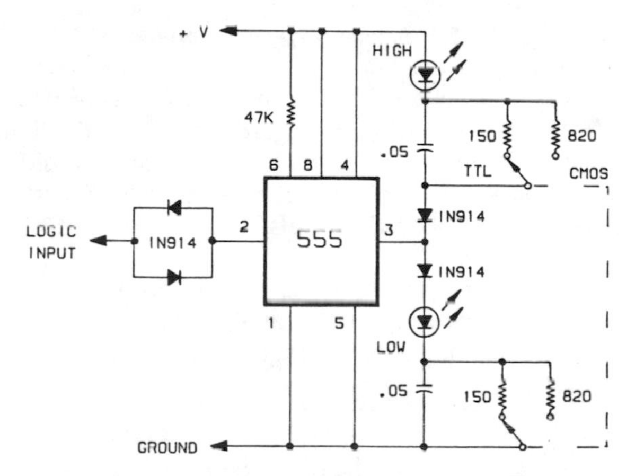

Reprinted from *Electronic Design*, June 7, 1976.

Fig. 5-4. An inexpensive logic probe using the 555 timer. (Copyright © 1976 Hayden Publishing Company, Inc.)

The capacitors across the series-limiting resistors pass current pulses to the LEDs during transitions in logic levels, and cause the LEDs to flash momentarily to indicate the presence of short pulses that would otherwise be undetected. In addition, the series diodes will protect the LEDs from excessive inverse voltages during capacitor discharge. As another convenient feature, the probe simply derives its operating power from the power supply of the circuit that is under test.

Another probe, shown in Fig. 5-5, is designed to indicate the logic states as well as pulses. The indicator system consists of three LEDs. A red LED lights up to indicate a logic 1, while a green LED lights

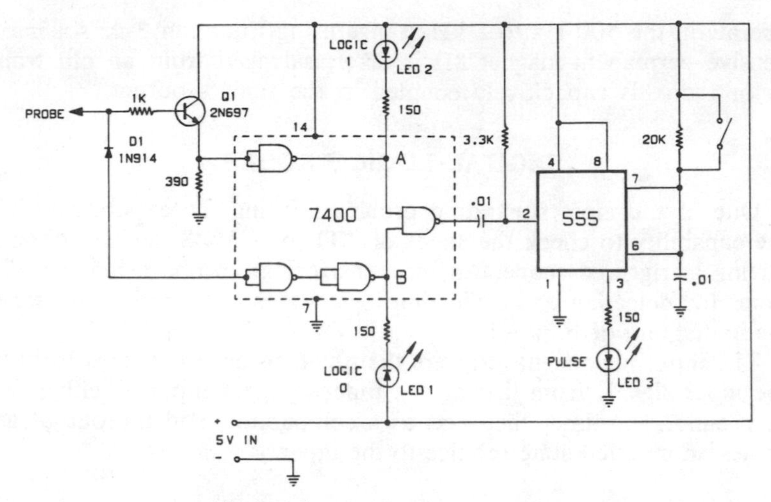

Fig. 5-5. A pulse/logic state probe.

for a logic 0. Also, a yellow LED comes ON for approximately 200 msec to indicate a pulse without regard to its width. This feature enables one to observe a short-duration pulse that would otherwise not be seen on the logic 1 and 0 LEDs. A switch across the 20 kΩ resistor can be used to keep this "pulse" LED ON permanently after a pulse occurs.

In operation, for a logic 0 input signal, both the 0 LED and the pulse LED will come ON, but the pulse LED will go OFF after 200 msec. For a logic 1 input, only the logic 1 LED will be lit. With the

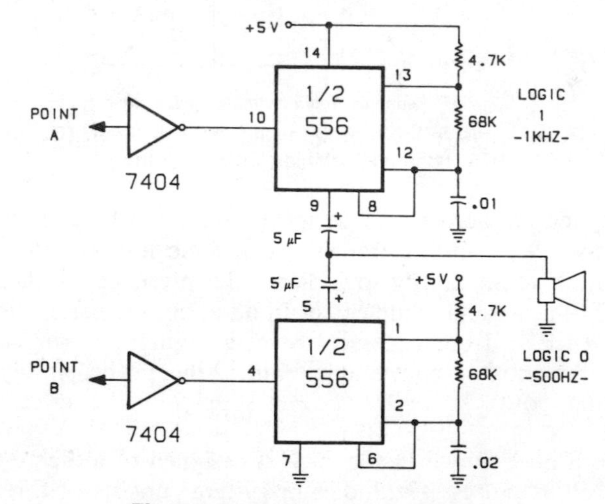

Fig. 5-6. Dual 555 audio oscillators.

switch closed, the circuit will indicate whether a negative-going or positive-going pulse has occurred. If the pulse is positive-going, both the 0 and pulse LEDs will remain ON. Conversely, the 1 and pulse LEDs will be ON to indicate a negative-going pulse.

Up to this point, we have neglected to consider the person who is handicapped because of color blindness or total blindness. Since colored LEDs may not be visible to some degree, these logic probes can be limited in their usefulness. A simple solution to this problem is to use two 555 audio oscillators, each with a different pitch to indicate the input logic states. For example, a logic 0 input might produce a 500-Hz tone, while a logic 1 input gives a 1-kHz tone. Using the 556 dual timer, the circuit of Fig. 5-6 can be added to the logic probe shown in Fig. 5-5.

MULTIWIRE CABLE TESTER

The convenient circuit, shown in Fig. 5-7, makes use of several 555 timers and single-package dual-color LEDs for the testing of multiwire cables. It will indicate which lines are open, shorted, or all right. Since each line is tested individually, faulty cables are quickly located and, therefore, easy to repair.

The circuit's clock is constructed with several 555 timers operating as a ring timer. As each timer turns ON in sequence, a positive pulse is applied to each of the lines under test. For a maximum of four lines, for example, five timers will be required.

The remaining section of the tester contains the LED indicators. For each line under test, there is a separate differential transistor pair driving a dual-color red/green LED. The red LED indicates a short and the green LED indicates an open line. Each differential pair looks for clock pulses at two places, namely at each end of the cable. If the same pulse is present at both ends of the same line, the differential transistor pair remains balanced and the LED for that line will not glow. However, if the clock pulse appears only at the "clock" end of the line, the differential pair is now unbalanced forcing current through the green LED, indicating that the line is open. Likewise, if a pulse appears only at the "indicator" end of the line, the differential pair "tilts" the other way causing a reverse current and the red LED indicates a short. When both LEDs remain dark, the cable line is all right.

The connectors on the cable lines can introduce another possible fault, namely, a short to the connector shell. This can be checked by the last timer. The timing-component values shown set the clock pulse width at 2 msec, which is fast enough to prevent lamp flicker, but not so fast as to cause capacitive coupling problems on long cables (up to 500 feet).

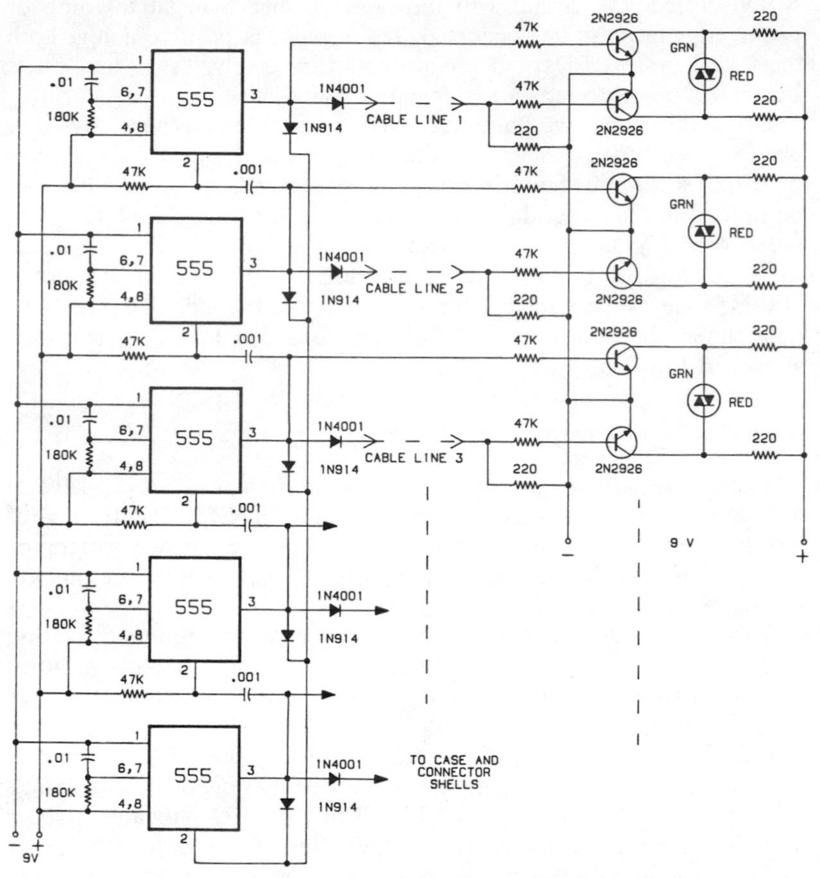

Reprinted from *Electronics*, May 10, 1973.

Fig. 5-7. A multiwire cable tester. (Copyright © McGraw-Hill, Inc., 1973.)

OSCILLOSCOPE DISPLAY

For low-cost oscilloscopes lacking a triggered sweep, the circuit of Fig. 5-8 should be a welcomed addition. When an input signal from the scope's vertical amplifier rises above the circuit's trigger-level voltage setting, the op amp switches, causing its output to go from $+V_{cc}$ to $-V_{cc}$. This voltage change is coupled to the timer's trigger input (pin 2) as a negative spike, setting its internal flip-flop and cutting off the discharge transistor.

The switch-selected timing capacitor C now charges exponentially through R4 until the capacitor voltage equals the timer's control voltage, $\frac{2}{3}$ V_{cc}. With the values shown, the trigger-sweep frequency can be varied from 1 Hz to 1 MHz.

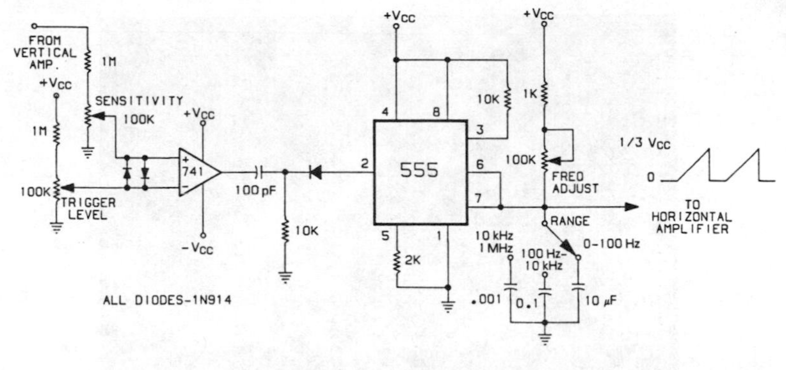

Reprinted from *Electronics*, October 11, 1973.

Fig. 5-8. Circuit for low-cost oscilloscopes lacking a triggered sweep. (Copyright © McGraw-Hill, Inc., 1973.)

Sometimes, when two or more analog signals, referenced to a common baseline, are simultaneously displayed on an oscilloscope, it is often difficult to identify each waveform. This can be especially true if the signals are less than 90° out of phase. By using a 555 timer, one can add dotted or dashed line markers to one or more analog signals.

Using the circuit of Fig. 5-9, the timer is connected as an astable multivibrator. The scope's vertical trace position at any instant of time is determined solely by the sum of the voltages across R4 and R5. When the square wave voltage across R4 is zero, the vertical position is determined solely by the analog input signal. When the square wave voltage across R4 is not zero, it drives the trace off the screen. This off-screen transition of the trace cannot be seen at normal scope intensity, as shown in Fig. 5-10.

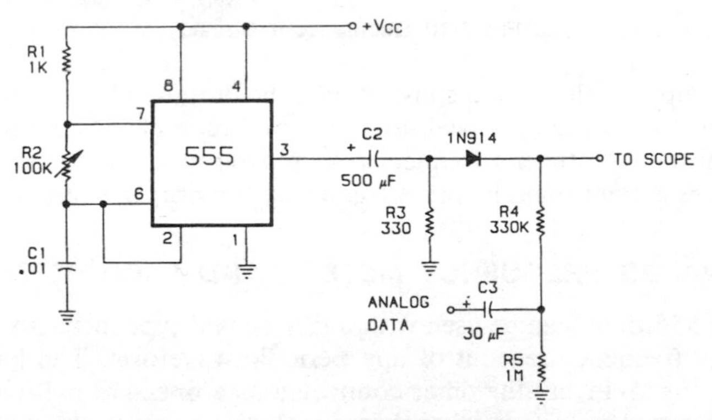

Reprinted from *Electronics*, April 29, 1976.

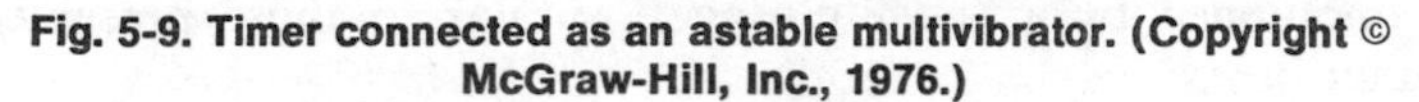

Fig. 5-9. Timer connected as an astable multivibrator. (Copyright © McGraw-Hill, Inc., 1976.)

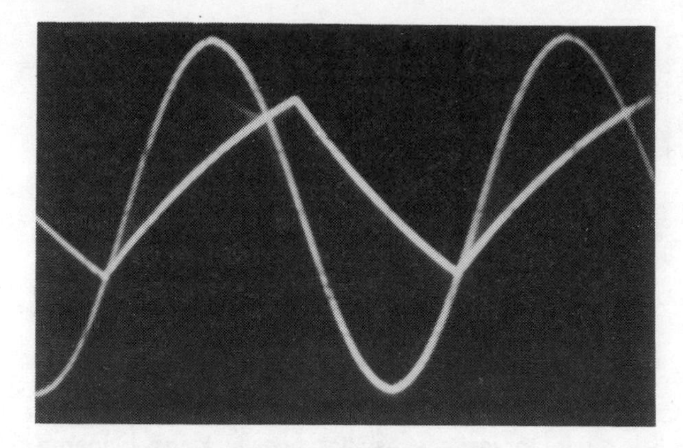

(A) Square wave voltage is zero.

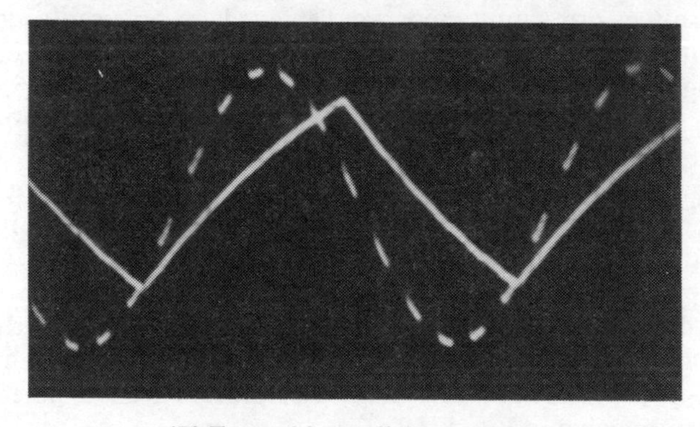

(B) Trace driven off the screen.

Fig. 5-10. Oscilloscope traces.

The supply voltage, and consequently the output voltage of the timer must be greater than the peak-to-peak amplitude of the analog signal. In addition, the timer's frequency should be adjusted to be at least ten times greater than the input frequency for proper operation.

ANALOG FREQUENCY METERS AND TACHOMETERS

The 555 timer can be used with a d'Arsonval type meter to provide a direct frequency readout of any periodic waveform. The basic circuit of Fig. 5-11 has the timer connected as a one-shot multivibrator. If the input waveform is other than a rectangular wave, the input must be conditioned by a Schmitt trigger, similar to those presented in Chapter 2.

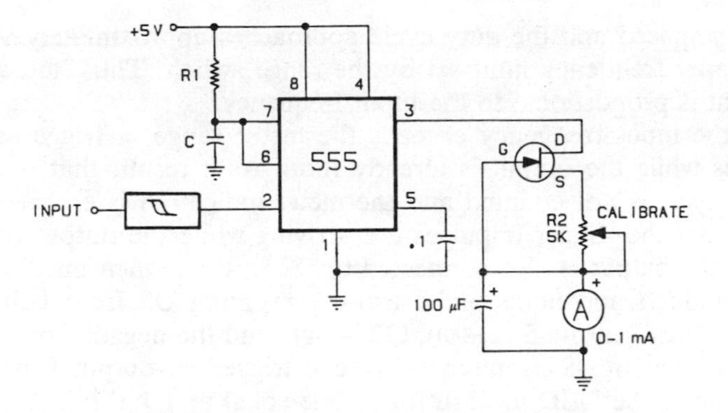

Fig. 5-11. 555 timer connected as a one-shot multivibrator.

The circuit is calibrated with R2 so that the full-scale frequency gives about a 40% duty cycle, which is then integrated or averaged by the meter to give a full-scale deflection. Frequencies less than full-scale change the ON/OFF ratio in a linear fashion so that the lower the frequency, the lower the proportionate reading. For instance, on the 1 kHz range, the full-scale period will be 1 msec. The timer is then set to give a 0.4-msec period by adjusting the R1C combination. The calibration control is then adjusted so that the average meter current is 1 mA, or a full-scale reading of "1000." The time period will be fairly independent of the supply voltage, but the applied meter voltage will not. This is the purpose of the FET which acts as a constant-current source in the meter line. It, therefore, provides a constant calibrated current, which is independent of the supply voltage.

A more versatile circuit, which uses the 556 dual timer, is shown in Fig. 5-12. In addition to providing readings from near dc to 50 kHz, it includes an over-range indicator. As before, one half of the dual timer (U1) is used as a monostable, triggered by the input frequency. In order to provide unambiguous measurements, the other half of the 556 uses another monostable to flash a warning light whenever the input frequency exceeds the maximum scale setting.

If the input is a rectangular pulse train, the pulses are then differentiated to produce the negative spikes that are needed to trigger the timer U1. For signals other than rectangular pulses, a Schmitt trigger must be used ahead of the circuit. When the trigger pin of U1 (pin 6) receives an incoming pulse, the output at pin 5 goes HIGH and delivers a current for a time period equal to 1.1 R1C1. This output pulse then appears only once for every cycle of the input. The current pulses, smoothed by C3, provide an average value that is indicated by the meter.

At low frequencies, the output pulses are well separated so that the average current is small. However, at higher frequencies, they are closely spaced and the duty cycle approaches approximately 95% at the upper frequency limit set by the range switch. Thus, the average current is proportional to the input frequency.

If the input frequency exceeds the meter range, a triggered spike arrives while the output is already HIGH. As a result, that particular input cycle is not counted and the meter indication is erroneous. To warn that the trigger impulses are arriving while the output of U1 is HIGH, the output is also connected to Q2 so that, when pin 5 is LOW, Q2 conducts and holds pin 8 HIGH, preventing U2 from being triggered. But when pin 5 is HIGH, Q2 is OFF, and the negative input spike that reaches pin 8 simultaneously can trigger an output from pin 9 that causes the LED to flash for a time equal to 1.1 R2C2.

For example, when the range switch is set to the 50-Hz range, any input frequency from near dc to 50 Hz causes the meter to read correctly, so that a frequency of 32 Hz produces a meter reading of 32 μA. On the other hand, frequencies greater than 50 Hz trigger the over-range circuit causing the LED to flash. Changing the switch to a higher range makes the LED stop flashing and the meter now reads correctly.

By essentially adding a photo-transistor, the basic circuit of Fig. 5-11 can be converted to operate as a photo-tachometer for measuring the rpm of any type of rotation element (up to 50,000 rpm). As

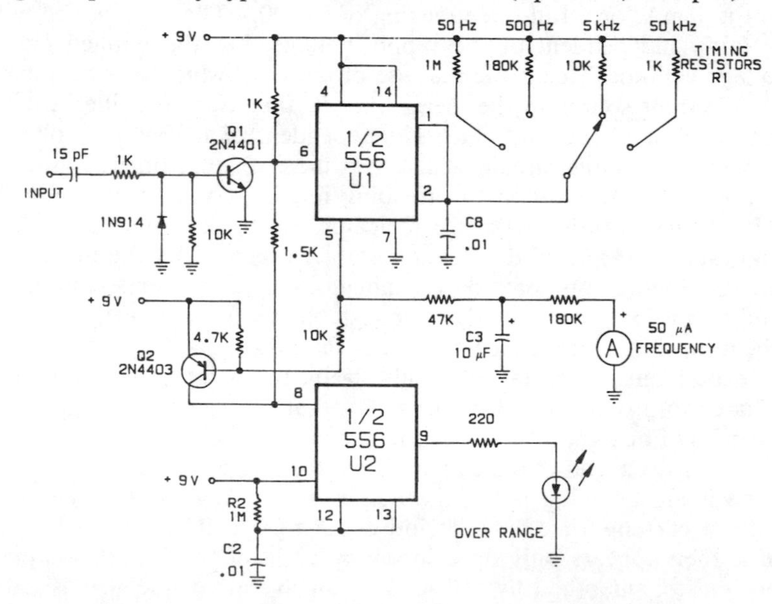

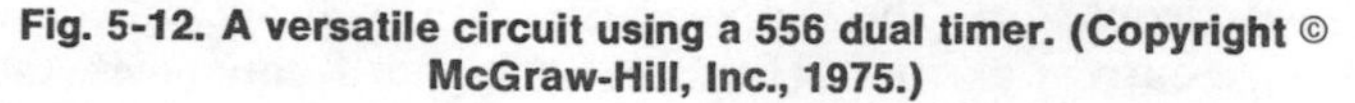
Reprinted from *Electronics*, April 17, 1975.

Fig. 5-12. A versatile circuit using a 556 dual timer. (Copyright © McGraw-Hill, Inc., 1975.)

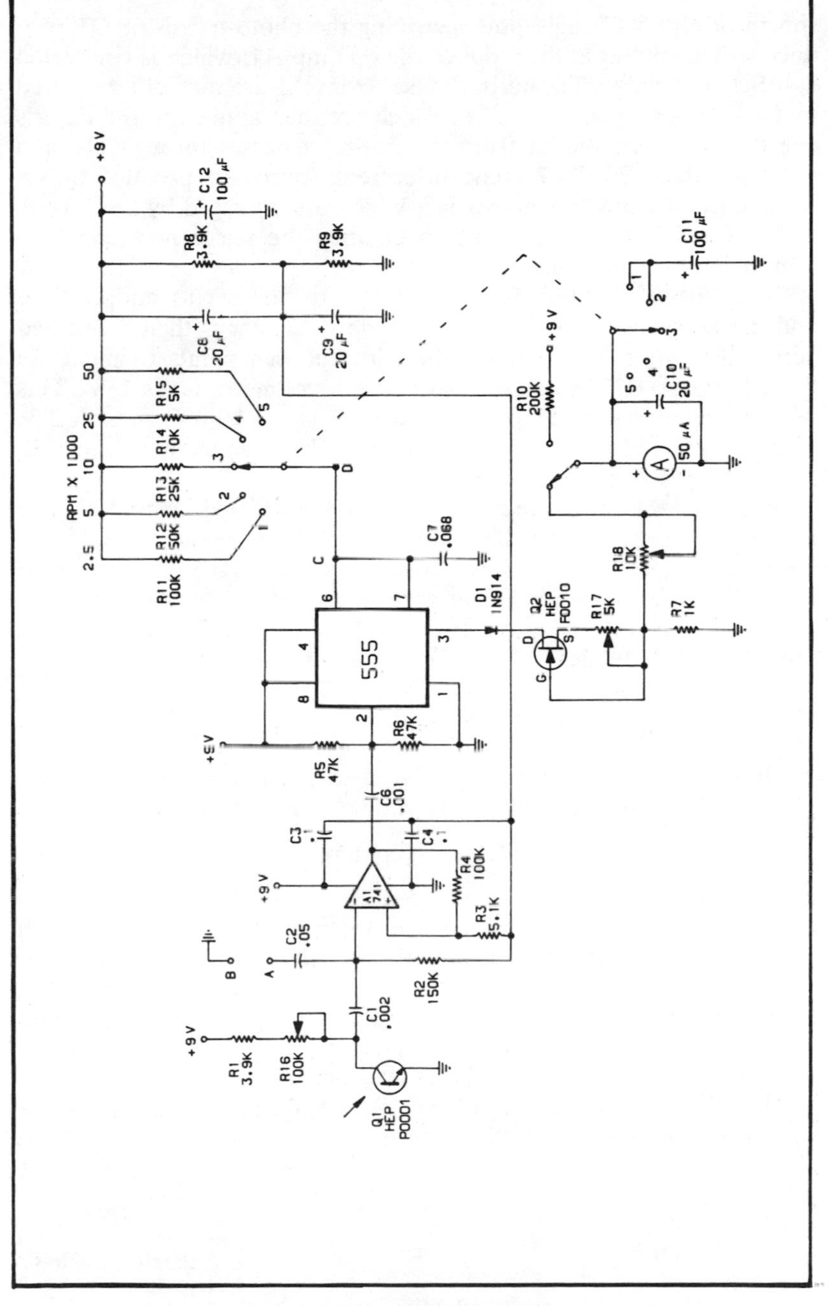

Fig. 5-13. Basic circuit in Fig. 5-11 converted to operate as a photo-tachometer.

shown in Fig. 5-13, light pulses striking the photo-transistor Q1 produce voltage pulses at the input of the op amp A1, which is connected as a Schmitt trigger. The output pulses from A1 are then differentiated by C6–R6, giving voltage spikes which are then applied to the timer's trigger input. The output from the one-shot passes through D1 and energizes the FET–R17 constant-current source to produce pulses with constant amplitude across R7, which are averaged by the 50 μA meter. Capacitor C11 is added to dampen the meter pointer vibration at the low rpm ranges.

To calibrate this unit, R17 and R18 are first set to midposition, and the range switch to 2500 rpm. A dc voltmeter is then connected across R7. After disconnecting the wire between points C and D in the schematic, R17 is adjusted so that the voltmeter reads 1 V. This wire is reconnected and the range switch is set to 10,000 rpm. A 3-V peak, 120-Hz, sine-wave signal is applied between points A and B, which is equivalent to 7200 rpm.

Finally, check for the rejection of low-level, 120-Hz modulation of incandescent light sources by aiming the photo-transistor at a 50- or 75-watt lamp while varying the sensitivity control R16 over its range. If the meter does not remain at zero under all conditions, the input hysteresis is increased by increasing R3 to 10 kΩ. Tachometers designed for automobiles will be covered in Chapter 7.

MEASUREMENT OF CAPACITANCE

From basic electronic theory, the charge between the capacitor's plates is the capacitance times the voltage across it,

$$Q = CV \quad \text{(coulombs)} \qquad \text{(Eq. 5-1)}$$

By using the simple circuit of Fig. 5-14, the unknown capacitor C_x is charged through R1 and D1 when the switch is open. When the switch is closed, C_x is discharged through the switch, D2, and the dc meter.

By using a 555 timer and a volt-ohm-millimeter, as shown in Fig. 5-15, the switch, power supply, and R1 are replaced with a square wave generator. When the timer's square wave is HIGH, C_x is charged to approximately 3.4 V and, when LOW, it discharges as before.

The 555 timer clock frequency (f) will determine the discharge current through the meter. Since, from Equation 5-1,

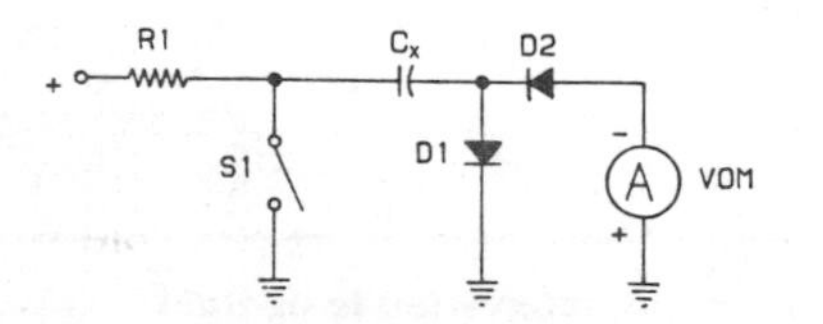

Fig. 5-14. A simple charging circuit.

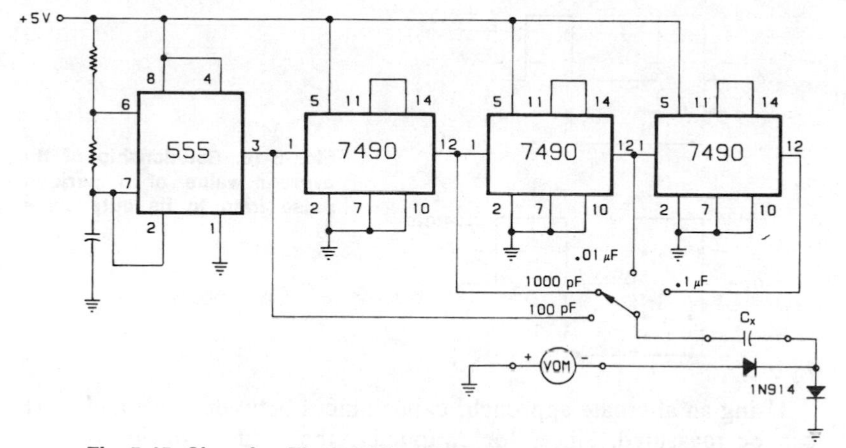

Fig. 5-15. Changing Fig. 5-14 by adding a square wave generator.

$$Q = C_xV$$

then,

$$fQ = fC_xV \qquad \text{(Eq. 5-2)}$$

However, the quantity fQ is the number of coulombs of charge stored per second, and is equal to the average current flowing through the capacitor,

$$fQ - \frac{dQ}{dt}$$

$$= i$$

or,

$$i = fC_xV \qquad \text{(Eq. 5-3)}$$

If, for example, the most sensitive scale of your VOM is 100 μA, the 555 clock frequency required to give a 100-pF full-scale reading is found by rearranging Equation 5-3,

$$f = \frac{i}{C_xV}$$

$$= \frac{100\ \mu A}{(100\ pF)\ (3.4\ V)}$$

$$= 294\ kHz$$

Larger values of capacitance can then be determined by decreasing the clock frequency, which is done digitally in decades by the type 7490 counters. Since the maximum oscillator frequency possible with the timer is about 300 kHz, this circuit is then limited to capacitances greater than 100 pF.

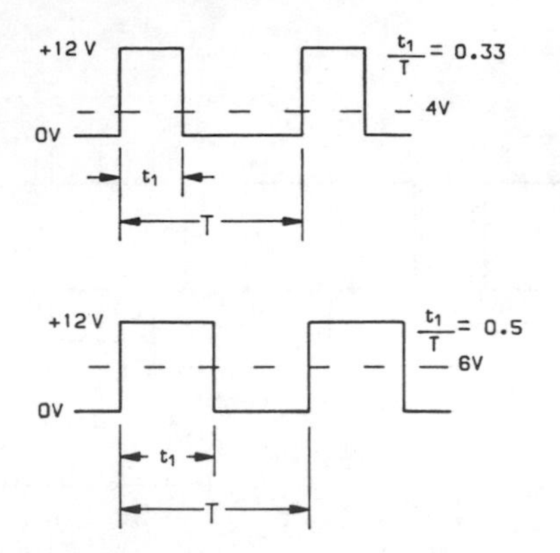

Fig. 5-16. Relationship of the average value of a periodic pulse train to its duty cycle.

Using an alternate approach, capacitances between 1 pF and 1 μF can be measured. Since, for simplicity, the readout device is a dc meter, it is required that there be a linear relationship between the unknown capacitance and the dc output of the measuring circuit. As illustrated in Fig. 5-16, the average or dc value of a periodic pulse train has a direct relationship to its duty cycle (Equation 4-2), or

$$\text{dc value} = V_p(t_1/T) \qquad \text{(Eq. 5-4)}$$

If the period T and the peak amplitude V_p of the pulse train can be held constant, the dc value will be directly proportional to the pulse width's HIGH time, t_1.

The block diagram of Fig. 5-17 illustrates how such a system would work. The trigger source is simply a free-running, constant-frequency pulse generator producing narrow negative output pulses. Each time a pulse occurs, the one-shot initiates an output pulse whose width t_1 is determined by the unknown capacitor C_x. Therefore, the larger the capacitor, the wider the pulse width. A dc meter then reads the average value of the pulse waveform. As pointed out in Chapter 2, the output pulse width must not exceed the time between triggers.

A circuit using a 555 timer and meeting the requirements outlined above is given in Fig. 5-18. The trigger source is a programmable unijunction transistor (PUT) Q1, and an inverter Q2. Since the out-

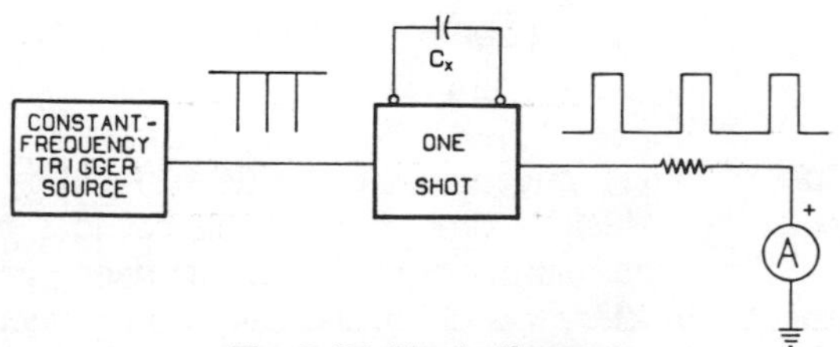

Fig. 5-17. Block diagram.

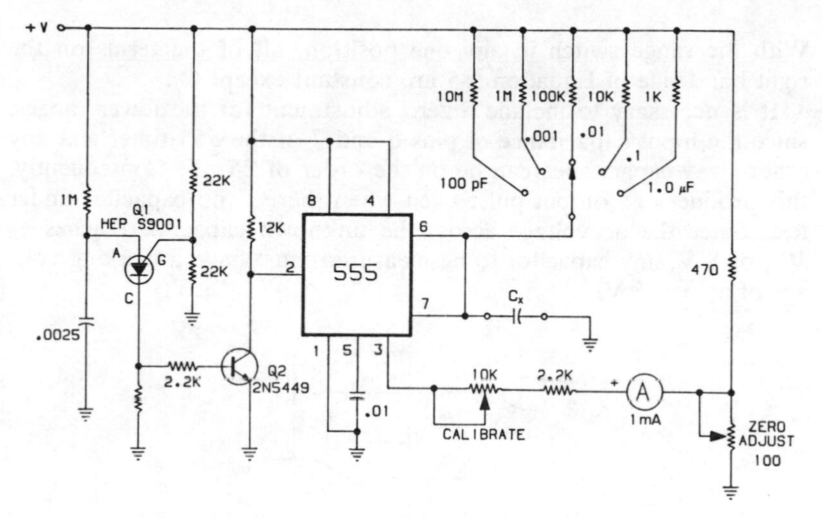

Reprinted by permission of *Ham Radio* magazine.

Fig. 5-18. Circuit for block diagram in Fig. 5-17. (Copyright © 1975 Communications Technology, Inc.)

put pulse width of the 555 timer equals 1.1 RC_x, Equation 5-4 can be rewritten as

$$\text{dc value} = \frac{1.1\ RC_xV_p}{T} \qquad \text{(Eq. 5-5)}$$

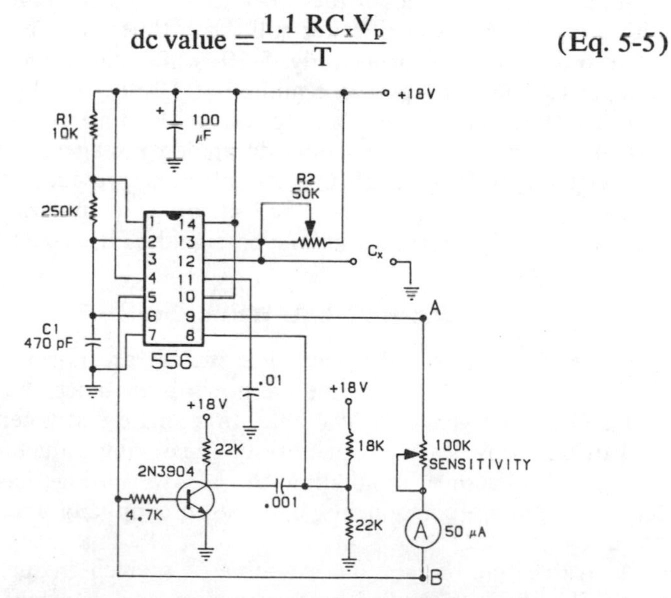

Reprinted from *Electronic Design*, March 1, 1976.

Fig. 5-19. The PUT and 555 timer of Fig. 5-18 replaced by a 556 dual timer. (Copyright © Hayden Publishing Company, Inc., 1976.)

With the range switch in any one position, all of the terms on the right hand side of Equation 5-5 are constant except C_x.

It is necessary to include a zero adjustment for the lower ranges, since the input capacitance of pins 6 and 7 of the 555 timer and any other stray capacitance can be on the order of 25 pF. Consequently, this produces an output pulse even when there is no capacitor under test. Since the dc voltage across the unknown capacitor reaches ⅔ V_{cc}, or 8 V, any capacitor to be measured must have a working voltage of at least 8 V.

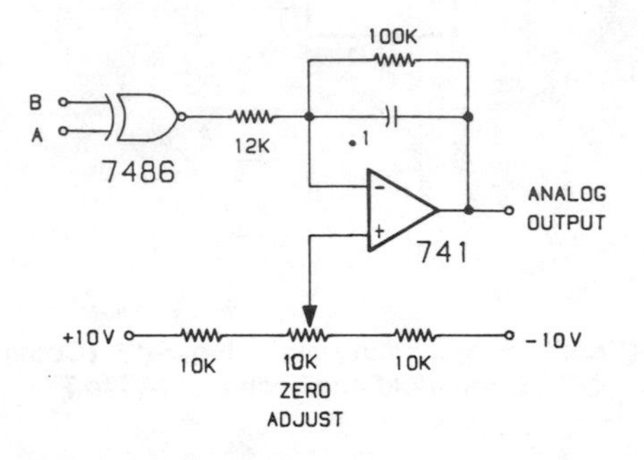

Fig. 5-20. Circuit designed to obtain a dc analog voltage from Fig. 5-19.

Since this book is about the 555 timer, the PUT can be replaced with another 555 device. Better still, both timers can be replaced by the 556 dual timer, shown in Fig. 5-19, which can detect capacitance changes as small as 1 part in a million at 10 nF. In addition, the circuit can detect changes over a range of 50 pF to 50 nF with the values shown. If desired, a proportional dc analog voltage can be obtained by using a type 7486 Exclusive-OR gate that provides the difference between the two pulse lengths, and an op amp integrator circuit to convert this difference into an analog signal (Fig. 5-20).

555-RLC MAXWELL BRIDGE

While ohmmeters and capacitance meters are normally available, the means of readily measuring unknown inductances between 1 μH and 1 H are not generally available. In a single instrument, the Maxwell bridge provides the capability of measuring inductance, capacitance, and resistance. In addition, the Maxwell bridge has the advantage of minimum components, calibration simplicity and, best of all, accuracy.

As with all ac bridges, an ac source of a given frequency must be used. This is where the 555 timer comes in, as illustrated in Fig. 5-21. Most of the other types of measuring bridges have more than one reactive element as compared to only one with the Maxwell. Over a

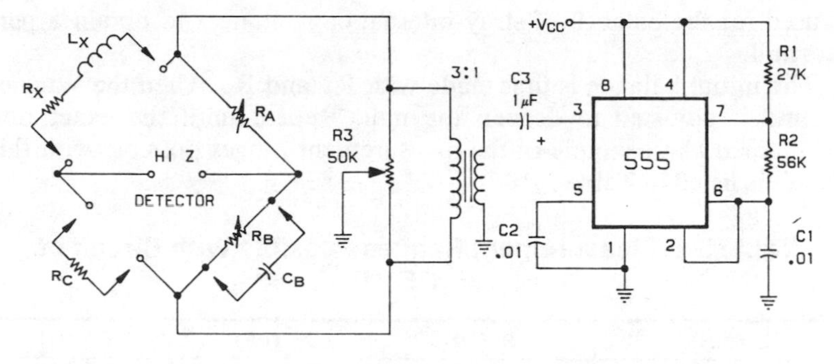

Fig. 5-21. A bridge circuit for measuring inductance. (Copyright © 1976 Communications Technology, Inc.)

fixed range, R_A can be calibrated to read inductance values directly, and at a specified frequency, R_C can be calibrated to read the Q, or quality factor.

Referring to Fig. 5-21, which is the arrangement for inductance measurements, and neglecting for the moment R3, the equation for bridge balance is then given by

$$L_x = R_A R_C C_B \qquad \text{(Eq. 5-6)}$$

where,

L is in mH,
R in kΩ,
C in μF.

For the series resistance of the unknown inductance,

$$R_x = \frac{R_A R_C}{R_B} \qquad \text{(Eq. 5-7)}$$

Then the Q factor is found from

$$Q = \frac{6.28\, fL_x}{R_x} \qquad \text{(Eq. 5-8)}$$

If a standard inductor is used in place of L_x, an unknown capacitor in place of C_B can be determined, so that

$$C_x = \frac{L_{standard}}{R_A R_C} \qquad \text{(Eq. 5-9)}$$

The 555 timer is wired as a 1-kHz clock which provides the necessary ac-signal source. The transformer, providing dc isolation, is a 3:1 (or less) audio type. Its secondary is connected to R3, whose rotor is grounded. This arrangement is called a Wagner ground, and is used for the balance of stray internal capacitances to obtain a perfect null.

The initial balance is first made with R_A and R_B. Then the Wagner ground is adjusted to deepen the null. Repeat until the exact null is obtained. An example of the measurement ranges possible with this circuit is listed in Table 5-1.

Table 5-1. Measurement Ranges Possible With Circuit of Fig. 5-21

L (mH)	R (Ω)	C_B (μF)	R_C (Ω)
0.01-0.1	1-10	0.01	10
0.1-1.0	10-100	0.01	100
0.1-1.0	1-10	0.1	10
1.0-10	10-100	0.1	100
1.0-10	1-10	1.0	10
10-100	100-1000	0.1	1000
10-100	10-100	1.0	100
100-1000	100-1000	1.0	1000

WAVEFORM GENERATORS

The simplest waveform generator that can be made with the 555 timer is an astable multivibrator, giving a rectangular output with variable frequency and duty cycle. Since this type of generator is fully discussed in Chapter 3, it will be omitted here. However, the 555 timer can be used to generate temperature-independent sawtooth waveforms whose linearity is within 1%, using the circuit shown in Fig. 5-22. By connecting the timer's trigger and threshold inputs together, it functions as the familiar astable. Capacitor C1 begins to

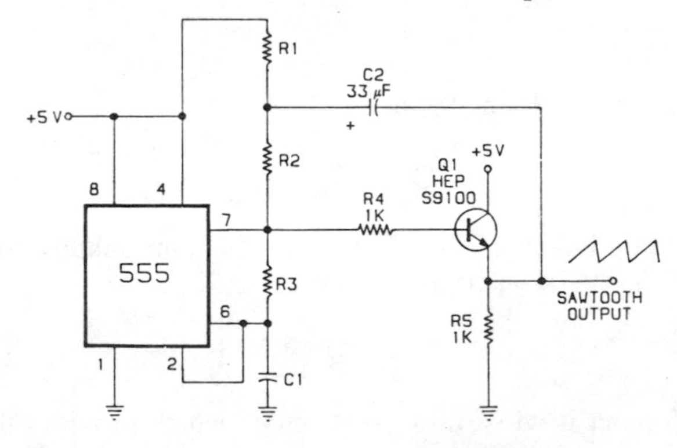

Reprinted from *Electronics,* March 18, 1976.

Fig. 5-22. Circuit for generating temperature-independent sawtooth waveforms. (Copyright © McGraw-Hill, Inc., 1976.)

charge through R1, R2, and R3 towards V_{cc}. For all practical purposes, the change in voltage at the junction of R2 and R3 is equal to that of pins 2 and 6. This change in voltage drives an emitter follower whose output is coupled back to the junction of R1 and R2. Thus, the voltage across R2 essentially remains constant during the charging cycle of C1, and produces the same effect of linear ramping as a constant-current source feeding C1.

Once this linear sawtooth signal at pin 6 reaches ⅔ V_{cc}, the internal comparator resets the timer's internal flip-flop, and the cycle is repeated. Resistance R3 is used to slow down the negative discharge-slope of the sawtooth. For proper operation, the following relationships apply:

$$R1 = R2$$
$$R2 \geqq 10\,R5$$
$$R3C1 \geqq 5\ \mu\text{sec}$$
$$R1C2 \geqq 10\,R2C2$$

so that the sawtooth frequency may be expressed as

$$f = \frac{1}{\{0.75(R1 + R2) + 0.693R3\}C1} \qquad \text{(Eq. 5-10)}$$

Although op amp circuits can easily generate a triangular wave by integration of a square wave, the peaks nevertheless become blunt at frequencies above 10 kHz unless expensive devices with high slew rates are used. Using the circuit of Fig. 5-23, triangular waveforms

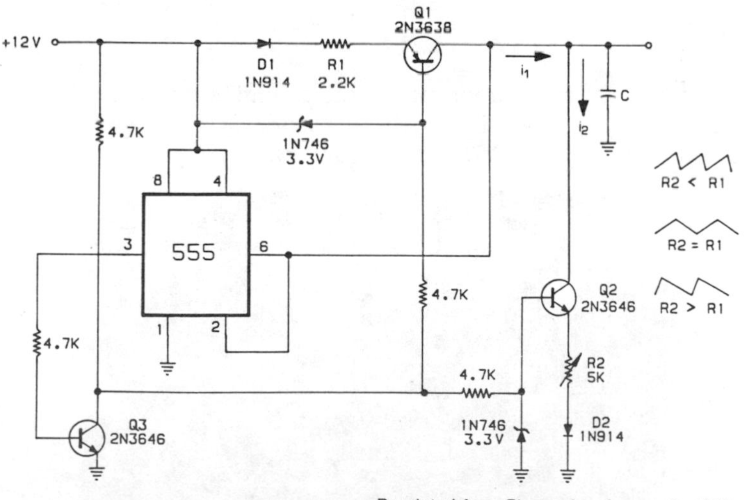

Reprinted from *Electronics*, January 8, 1976.

Fig. 5-23. Circuit for generating triangular waveforms. (Copyright © McGraw-Hill, Inc., 1976.)

with adjustable symmetry can be generated by alternately charging and discharging a capacitor.

Transistors Q1 and Q2, with their corresponding zener diodes, act as a switched-current source and sink, activated by Q3. When Q3 is ON, Q1 is switched ON and a current i_1 charges capacitor C. The linear voltage ramp that appears across capacitor C corresponds to the relationship

$$\frac{dV_c}{dt} = \frac{i_1}{C} \qquad \text{(Eq. 5-11)}$$

The voltage across C then increases until $V_c = \frac{2}{3} V_{cc}$, at which point the output of the timer goes LOW, turning OFF Q3. The Q1 current source is then deactivated while the Q2 current sink is switched ON. The capacitor is now discharged by current i_2 until $V_c = \frac{1}{3} V_{cc}$ and the cycle starts over again. For a 12-V supply, the output voltage varies between 4 V and 8 V. For the values shown, the triangle-wave frequency is approximated by

$$f \simeq \frac{75}{C} \qquad \text{(Eq. 5-12)}$$

Resistor R2 is a symmetry adjustment, controlling the capacitor discharge rate by varying i_2, as shown in Fig. 5-23.

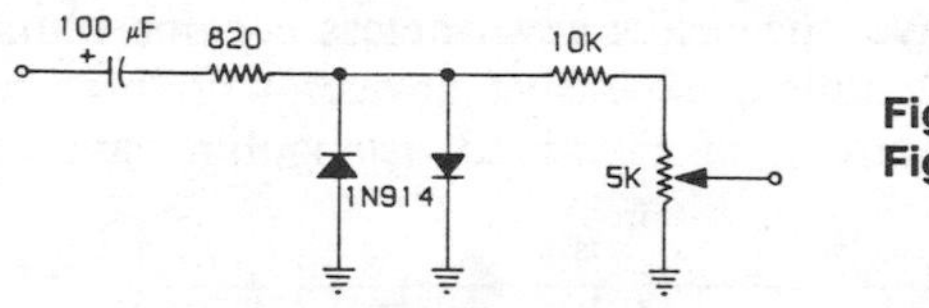

Fig. 5-24. Circuit addition to Fig. 5-23 necessary to obtain a sine wave.

By adding several components to the triangular-wave generator, as shown in Fig. 5-24, we can obtain a reasonably good approximation of a sine wave (Fig. 5-25). The diodes play a dual role. They clip

Fig. 5-25. Output from Fig. 5-24.

the triangular peaks and they also set the clipping level. Uneven clipping can be improved by trying other diodes, since some diodes will generate a better sine wave approximation than others. If necessary, a lower output impedance of about 1 Ω for the diode circuit can be obtained by using an op amp voltage follower after the 5 kΩ potentiometer.

TEMPERATURE MEASUREMENT AND CONTROL

By now you may have come to the conclusion that the 555 timer is almost as versatile as the op amp. Another of the timer's many applications is the ability to measure and control temperature using a thermistor. When wired as an astable, the 555 can be used to generate a square-wave output voltage whose frequency has a one-to-one correspondence with temperature.

As we already know, the conventional astable circuit for the 555 has two fixed resistors. As shown in Fig. 5-26, a negative temperature-coefficient thermistor R_T (in series with a fixed resistor) replaces one of these, while the other fixed resistor is replaced by a transistor which is turned ON during the capacitor's charging cycle. The near-zero ON resistance and very large OFF resistance of transistor Q1 yields equal charge and discharge periods that depend only on RT and RS. Then the operating frequency is the same as given by Equation 3-9, except that R1 has been replaced by the quantity (RT + RS), or

$$f = \frac{0.722}{(RT + RS)C} \qquad \text{(Eq. 5-13)}$$

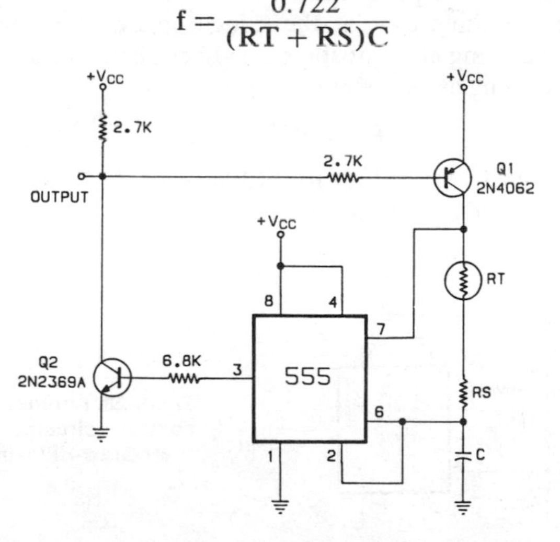

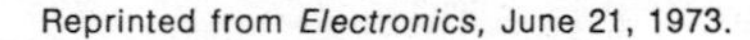
Reprinted from *Electronics*, June 21, 1973.

Fig. 5-26. Using a thermistor in a 555 astable circuit. (Copyright © McGraw-Hill, Inc., 1973.)

Since all the operating voltages for the 555 timer are derived as ratios of V_{cc}, the internal trigger-comparator can then be used with a thermistor-resistor divider network to form the temperature controller of Fig. 5-27.

When the thermistor RT cools below a set value, the voltage at pin 2 drops below $\frac{1}{3}$ V_{cc}, turning ON the triac-controlled heating element

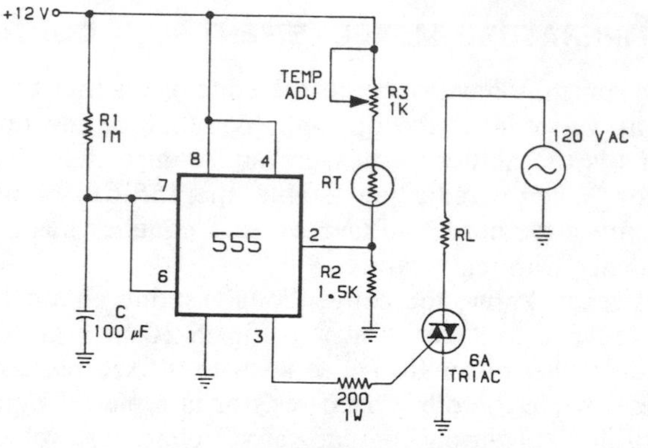

Reprinted from *Electronic Design*, August 16, 1975.

Fig. 5-27. A temperature-controller circuit. (Copyright © Hayden Publishing Company, Inc., 1975.)

and starting the timing cycle. If the thermistor temperature rises above the predetermined set point before the end of the timing cycle, the heating element shuts OFF at the conclusion of the timing period; otherwise the heating element stays ON. Thermistors of different values can be used as long as the relationship

$$R3 + RT = 2\ R2 \qquad \text{(Eq. 5-14)}$$

holds. Although larger values for R3 provide wider adjustment, the sensitivity is nevertheless reduced.

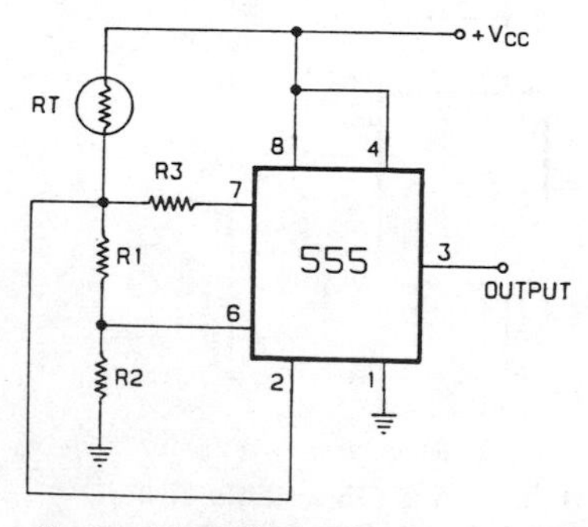

Reprinted from *Electronics*, June 21, 1973.

Fig. 5-28. Another temperature-control circuit. (Copyright © McGraw-Hill, Inc., 1973.)

Another circuit that lends itself to temperature-control applications, particularly those reserved for thermostats that must maintain an environment within a given temperature range, is shown in Fig. 5-28. For the thermostat, the thermistor-resistor divider networks produce a voltage that is directly proportional to temperature. When the temperature is rising, the output of the timer is HIGH, and the threshold input voltage is determined by the voltage divider set up by RT, R1 and R2, which increases as RT decreases. When RT equals the thermistor resistance at the "hot" setpoint temperature, R_{TH}, the divider relationship needed to establish a voltage of ⅔ V_{cc} at the threshold input is then

$$\frac{R_{TH} + R1}{R_{TH} + R1 + R2} = 0.5 \qquad \text{(Eq. 5-15)}$$

After an input to the internal comparator reaches this level, the discharge transistor is switched ON, effectively placing R3 in parallel with (R1 + R2). As the temperature drops, RT increases so that the voltage is divided between RT and R3 in parallel with (R1 + R2). When RT equals the resistance at the "cold" setpoint temperature, R_{TC}, the divider produces a voltage of ⅓ V_{cc} at pin 2. Then, the divider relationship becomes

$$0.5 = \frac{R3 \parallel (R1 + R2)}{R_{TC} + [R3 \parallel (R1 + R2)]} \qquad \text{(Eq. 5-16)}$$

where,

$$R3 \parallel (R1 + R2) = \frac{R3(R1 + R2)}{R1 + R2 + R3}$$

When a standard thermistor is used, and its resistance as a function of temperature is known, we can determine the required values of R1, R2, and R3 based upon the ratio $R_{TC}/R_{TH} = \alpha$. If $\alpha \geqq 2$, then

$$R1 = (0.5\alpha - 1)R_{TH} \qquad \text{(Eq. 5-17)}$$

$$R2 = \alpha R_{TH} \qquad \text{(Eq. 5-18)}$$

$$R3 = (3\alpha^2 - 1)R_{TH}/(4\alpha - 2) \qquad \text{(Eq. 5-19)}$$

However, if $\alpha < 2$,

$$R1 = 0 \qquad \text{(Eq. 5-20)}$$

$$R2 = 2R_{TH} \qquad \text{(Eq. 5-21)}$$

$$R3 = 2R_{TH}R_{TC}/(2R_{TH} - R_{TC}) \qquad \text{(Eq. 5-22)}$$

To prevent noise signals from triggering the timer prematurely, pins 2 and 6 should be bypassed with 0.01 μF disc capacitors.

ADDITIONAL CONTROL CIRCUITS

The brightness of single or 7-segment LED displays can be controlled by simply changing the supply voltage. However, this has the disadvantage of wasting power. A more efficient solution is to turn the display OFF and ON very rapidly, adjusting the ratio of the OFF time to the ON time to obtain the desired duty cycle. As has been pointed out before, the average voltage and, consequently, the average brightness of the LED display will depend on the duty cycle with no power dissipation during the OFF periods (Equation 4-2).

Fig. 5-29 shows how a 555 timer can be used to go from a duty cycle of 50% to 99%. By using the LOW period of the output to unblank or power the display, and setting a 50% duty cycle as the maximum brightness, one can control the display's brightness from virtually OFF to maximum.

On the other hand, the brightness of a LED display can be varied automatically by combining cadmium-sulfide photocell with a 555 timer as a pulse width modulated astable. The circuit of Fig. 5-30 shows the usual astable configuration, but the photocell replaces one of the timing resistors, and diode D1 bypasses the other timing resistor during the charging of the capacitor. Consequently, the maximum duty cycle is increased beyond the normal 50%, allowing the display to obtain full brightness.

As the increasing ambient light intensity decreases the photocell's resistance, the timer's duty cycle increases. As before, the varying duty cycle controls the length of time the display is ON, which is the same as controlling the brightness.

The 555 can also serve as a bistable switch controlled by a pair of photocells as shown in Fig. 5-31. When activated by external light sources, the circuit can be used for a variety of remote control applications, such as turning on the coffee pot with the rays from the rising sun.

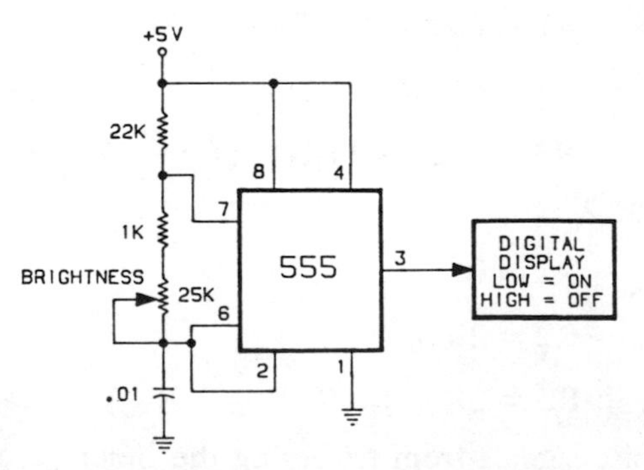

Fig. 5-29. LED brightness control circuit.

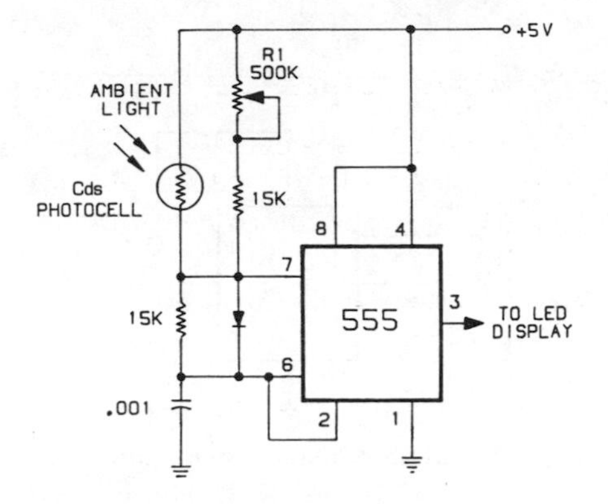

Reprinted from *Electronics*, December 26, 1974.

Fig. 5-30. Using a photocell and a 555 timer to automatically vary the brightness of a LED display. (Copyright © McGraw-Hill, Inc., 1974.)

In operation, light striking one of the photocells will switch the 555 from a HIGH to a LOW state, or vice versa. If HIGH initially, light striking PC1 causes a positive input pulse that switches the timer which activates an optocoupler, isolating the 555 from the ac power line. This in turn gates the triac, switching it ON and applying ac power to the load. If the timer is LOW, light striking PC2 will do the reverse. Resistors R1 and R2 balance the photocell characteristics. However, if the photocells are already closely matched, R1 and R2 can be omitted.

Another bistable configuration for the timer is shown in Fig. 5-32, which can function as a bistable flip-flop for TTL drivers, displays, or latch elements. The trigger input is an active LOW "set" function. Pin 4 then serves as an active LOW reset, and the threshold pin is the active HIGH reset. Both resets can be used, or just one, with the other

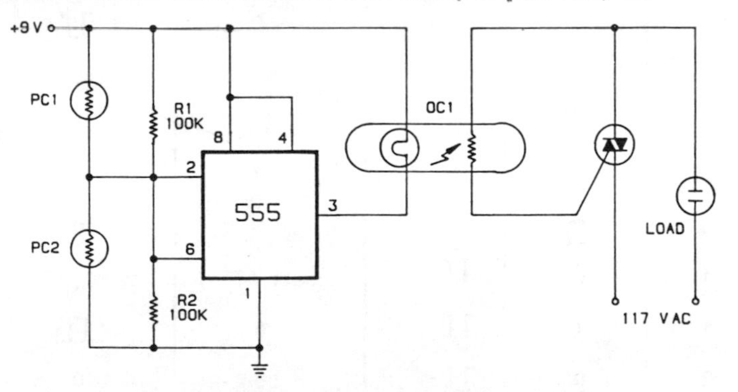

Reprinted by permission of *Popular Electronics* magazine.

Fig. 5-31. The 555 timer connected as a bistable switch. (Copyright © 1976 Ziff-Davis Publishing Company.)

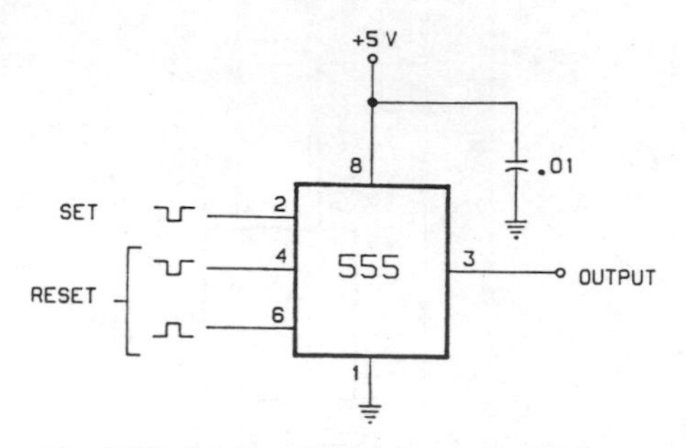

Fig. 5-32. Another bistable configuration.

connected in its inactive state. The listing in Table 5-2 shows how the output responds to various inputs. However, as stated in Chapter 1, the internal 555 timer circuitry differs slightly from manufacturer to

Table 5-2. How Output of Circuit in Fig. 5-32 Responds to Various Inputs

INPUTS			OUTPUT	
Pin 4 (reset) (active LOW)	Pin 6 (threshold) (active HIGH)	Pin 2 (trigger) (active LOW)	National LM555H	Signetics NE555V
⊔	0	1	resets (↓)	resets (↓)
⊔	1	1	0	0
⊔	0	0	⊔	⊔
⊔	1	0	0	⊔
1	⊓	1	resets	resets
1	⊓	0	⊔	1
0	⊓	1	0	0
0	⊓	0	0	0
1	0	⊔	sets (↑)	sets (↑)
1	1	⊔	0	⊓
0	0	⊔	0	0
0	1	⊔	0	0

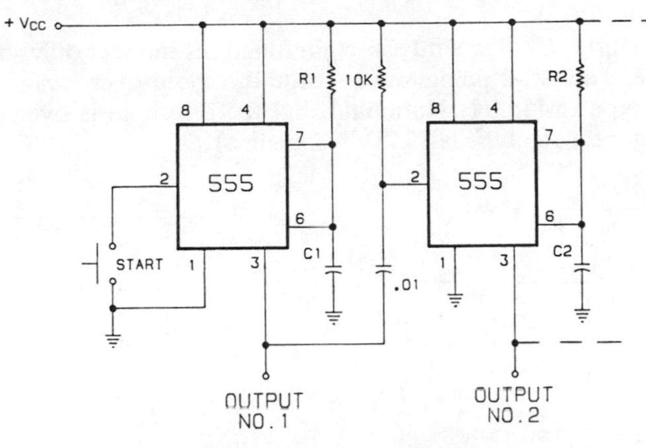

Fig. 5-33. A sequence generator.

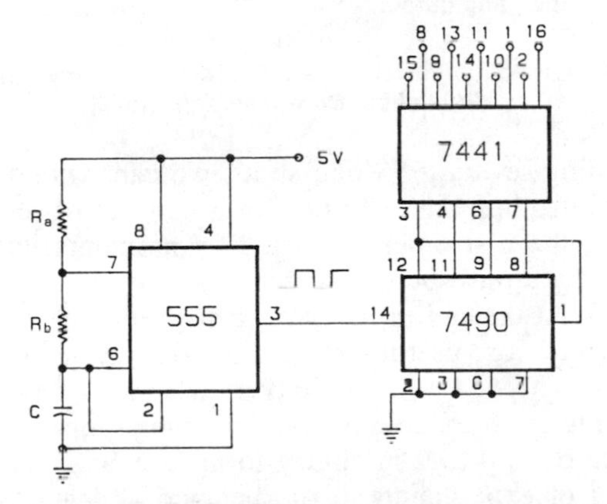

Reprinted by permission of *Popular Electronics* magazine.

Fig. 5-34. A useful circuit that uses one 555 timer. (Copyright © 1976 Ziff-Davis Publishing Company.)

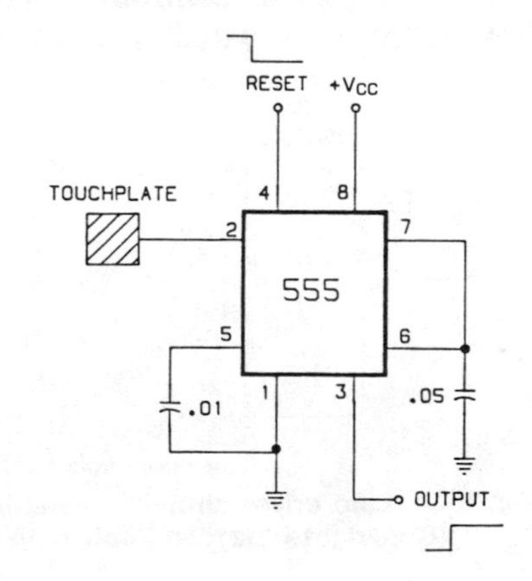

Fig. 5-35. A simple touch-control circuit.

manufacturer. Consequently, certain functions interact differently. For example, Table 5-2 points out that the threshold overrides the trigger for the type LM555H (National), but the threshold is overridden by the trigger for the type NE555V (Signetics).

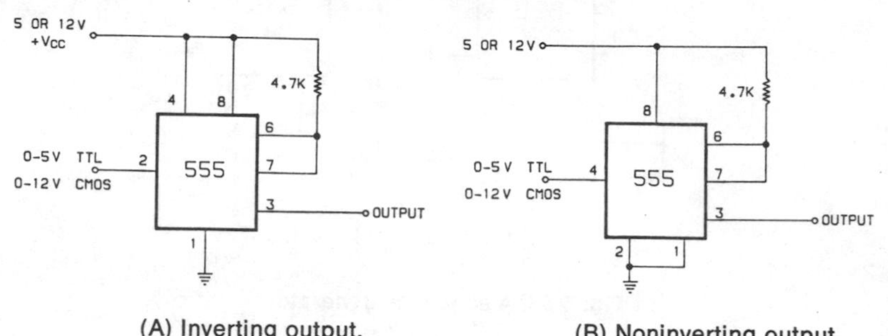

(A) Inverting output. (B) Noninverting output.

Reprinted from *Electronic Design,* January 19, 1976.

Fig. 5-36. A medium-current line driver circuit. (Copyright © Hayden Publishing Company, Inc., 1976.)

By cascading several 555 one-shots, we can have a sequence generator for controlling the operation of a number of sequential events. In Fig. 5-33, the first timer is started by momentarily grounding the trigger input by a push-button switch, and the timer runs for a period of 1.1 R1C1. At the end of its cycle, it triggers the next timer, and so on. However, a more useful circuit using only a single timer is shown in Fig. 5-34. The 555 astable delivers a series of pulses to a 7490 decade counter, which counts the incoming pulses and supplies a running total from 0 to 9 in binary form to a BCD/decimal decoder.

A number of experiments in the behavioral sciences study the reaction times of laboratory animals by starting and stopping some kind of timing circuit, such as a digital clock. A simple circuit that can be used for "touch" control is shown in Fig. 5-35. It is simply a one-shot multivibrator without the timing resistor. Touching a metal

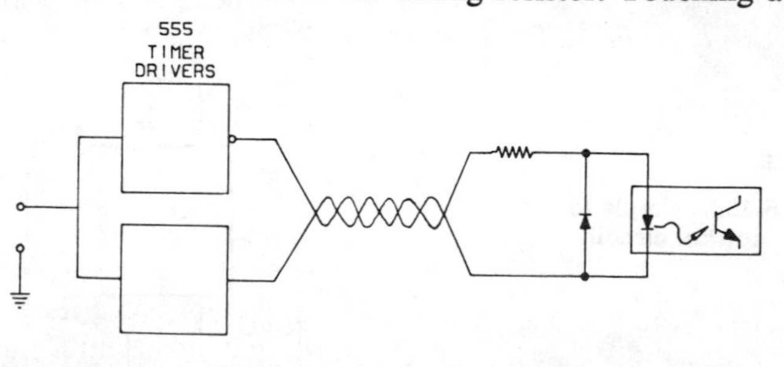

Reprinted from *Electronic Design,* January 19, 1976.

Fig. 5-37. Line driver circuit terminating with an optical isolator. (Copyright © Hayden Publishing Company, Inc., 1976)

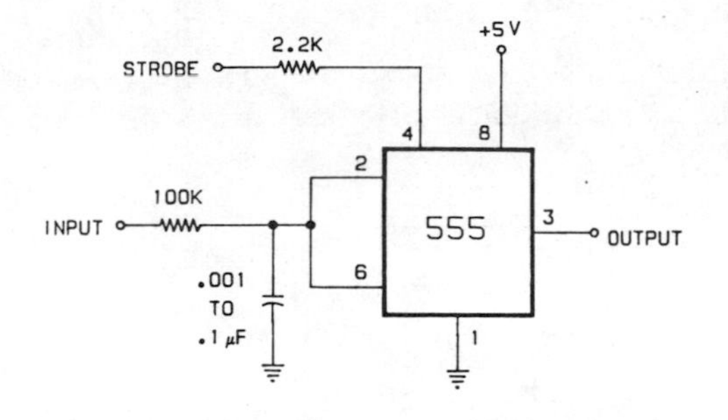

Reprinted from *Electronics*, June 21, 1973.

Fig. 5-38. A level-sensing 555 timer circuit. (Copyright © McGraw-Hill, Inc., 1973.)

plate triggers the timer, but since there is no external timing resistor, the output is latched HIGH indefinitely. The circuit is reset by application of a negative pulse at pin 4.

Oddly enough, a 555 timer can be used as a medium-current line driver with either an inverting or noninverting output, which can either source or sink up to 150 mA (Fig. 5-36). This drive capacity is adequate for either single ended or balanced lines that terminate with optical isolators, as shown in Fig. 5-37.

Since the 555 contains a comparator internally, it can simultaneously serve as a level shifter. For example, with only a 5-V input signal, a 12-V swing can be applied to the line.

On the other hand, the 555 timer can also be used as a level-sensing device when preceded by an RC integrator to form a noise-immune line receiver. The receiver must have a high input impedance and must not require a special driver at the sending end. Using the circuit of Fig. 5-38, only one signal conductor is required and it can be unshielded.

CHAPTER 6

Playing Games With the 555 Timer

In this chapter, several circuits will be presented that will help you make decisions, test your sense of timing, or arouse your competitive spirit.

COIN FLIP DECISION MAKER

The simple electronic coin flip (heads or tails) circuit of Fig. 6-1 can be also used as a "high level" decision maker to provide a YES or NO answer. In operation, the 555 timer is wired as an astable oscillator, driving a type 7473 TTL flip-flop. The flip-flop then alternately turns ON and OFF a pair of LEDs. When the "decision" switch is pressed, the LEDs flash at a 2 kHz rate, which is faster than the eye can follow, so that they both appear to be ON. At any desired time, the switch is released and only one LED will be ON. This selection is then essentially a random choice depending on the precise instant that the switch is released.

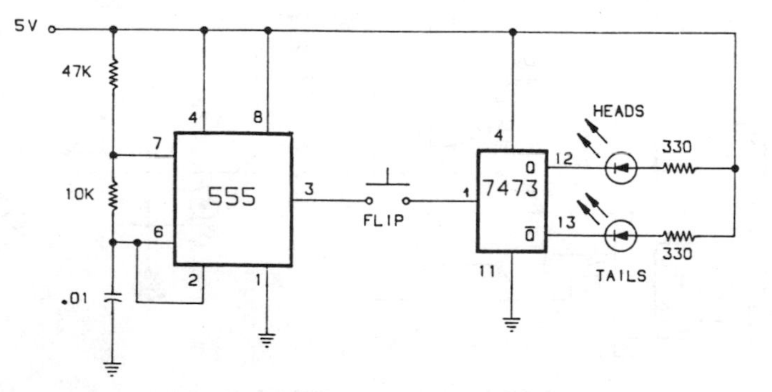

Fig. 6-1. Electronic coin flip circuit.

CASINO GAMES

To enjoy the atmosphere of casino-style gambling, you would have to travel to such places as Las Vegas, London, Macau, or Monte Carlo, to name a few. However, we will discuss a few circuits that will bid for the favors of Lady Luck at home.

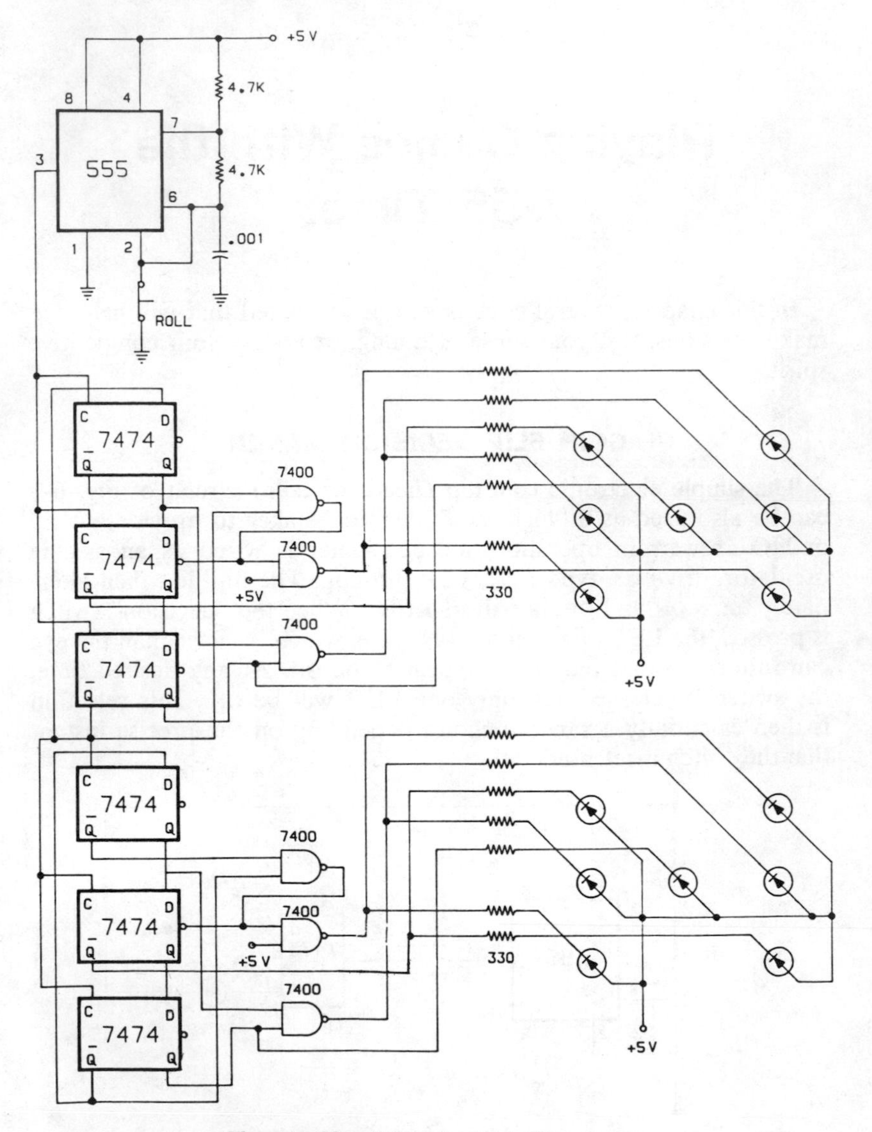

Fig. 6-2. Circuit for electronic dice game.

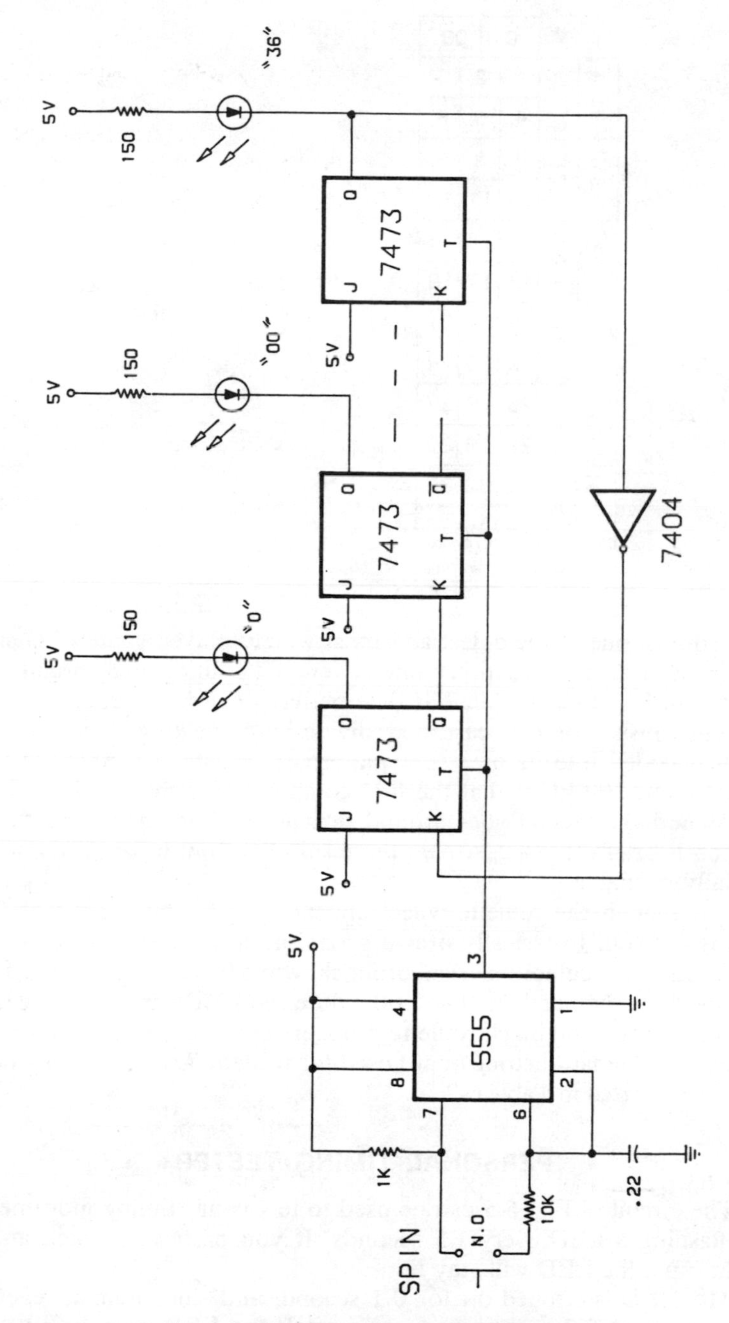

Fig. 6-3. Circuit for electronic roulette.

		0		00
1–18	1 ST 12	1	2	3
		4	5	6
EVEN		7	8	9
		10	11	12
RED	2 ND 12	13	14	15
		16	17	18
BLK		19	20	21
		22	23	24
ODD	3 RD 12	25	26	27
		28	29	30
19–36		31	32	33
		34	35	36
		2 TO 1	2 TO 1	2 TO 1

Fig. 6-4. Standard roulette betting board.

Perhaps, one of the oldest and most widely played game of chance is Craps. Using a 555 timer and a few TTL integrated circuits, as shown in Fig. 6-2, the familiar dice patterns can be reproduced. The circuit consists of two cascaded divide-by-6 walking-ring counters, each decoded into its own spot pattern. The astable circuit runs so fast (about 100 kHz) that the first counter cycles many hundreds of times and the second goes around dozens of times while the "roll" button is briefly pressed, so that the results turn out to be random and equally probable.

However, if the roulette wheel appeals to you, the circuit of Fig. 6-3 is for you. It basically uses a series of shift registers to drive 38 LEDs in the circular numerical order shown in Table 6-1. Green LEDs are used for the numbers 0 and 00, while red LEDs and yellow LEDs are for the red and black roulette numbers respectively. Fig. 6-4 illustrates the standard betting board used for roulette and the betting odds are summarized in Table 6-2.

PERSONAL TIMING TESTER

The circuit of Fig. 6-5 can be used to test your "timing judgement" by flashing a LED every 1.5 seconds. If you press the switch at the right time, the LED will stay lit.

The LED is strobed ON for 0.1 second, and since human reaction time is about 0.3 second, you can't catch the LED once it is ON. It

Table 6-1. Circular Numerical Lighting Order of LEDs

Clockwise		
	1 (R)	2 (B)
	13 (B)	14 (R)
	36 (R)	35 (B)
	24 (B)	23 (R)
	3 (R)	4 (B)
	15 (B)	16 (R)
	34 (R)	33 (B)
	22 (B)	21 (R)
	5 (R)	6 (B)
	17 (B)	18 (R)
	32 (R)	31 (B)
	20 (B)	19 (R)
	7 (R)	8 (B)
	11 (B)	12 (R)
	30 (R)	29 (B)
	26 (B)	25 (R)
	9 (R)	10 (B)
	28 (B)	27 (R)
	0 (G)	00 (G)

R = Red; B = Black; G = Green

will then be necessary to judge the time which has passed after the LED turns OFF before pressing the switch.

The circuit uses a 555 timer as an astable multivibrator. Since the switch is normally closed, C1 starts to charge up through R1 and R2. When the voltage across C1 equals ⅔ V_{cc}, the timer's output changes state, allowing current to flow through the LED. When the timer's output is HIGH, no current flows through the LED. Consequently, the

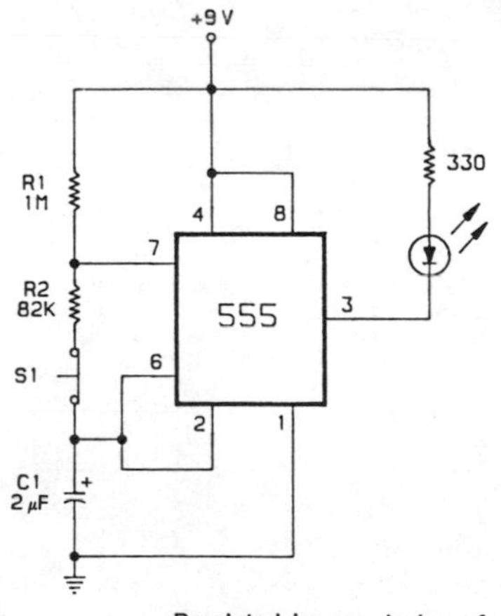

Fig. 6-5. A circuit for testing your judgement of time. (Copyright © 1975 Ziff-Davis Publishing Company.)

Reprinted by permission of *Popular Electronics* magazine.

Table 6-2. Roulette Betting Odds

Type of Bet	Odds
Red or Black	Even
Odd or Even	Even
High–Low: 1–18 or 19–36	Even
Dozen: 1st 12, 2nd 12, or 3rd 12	2 to 1
Columns: 1–34, 2–35, or 3–36	2 to 1
Six Line: 1–6, 7–12, 13–18, etc.	5 to 1
Five Line: 1, 2, 3, 0, and 00	6 to 1
Corners: In cross between 4 numbers, e.g. 5, 6, 8, 9	8 to 1
Rows: 1–3, 4–6, 7–9, etc.	11 to 1
Split: Between 2 adjacent numbers, or 0–00	17 to 1
Straight Up: Any single number	35 to 1

LED glows only during the discharge cycle. Since the capacitor charges through R1 and R2 but discharges only through R2, the discharge time period is much less than the charge time (see Equations 3-2 and 3-3).

If the switch is pressed at any time during the cycle, the charge and discharge paths are immediately opened. Since the voltage across C1 remains fixed, the output remains in the same state as when the switch was pressed. Thus, if the switch is opened while the LED is ON, it will stay lit. Closing the switch allows the cycle to resume at the point where it was interrupted.

CHAPTER 7

Circuits for the Automobile and Home

This chapter contains some warning-type and often helpful circuits for use on your automobile. Then a toxic gas alarm and a weather monitor are discussed.

WINDSHIELD WIPER SYSTEM

A simple audible monitor to warn the driver when the fluid is getting low in the windshield-washer system is shown in Fig. 7-1. Two

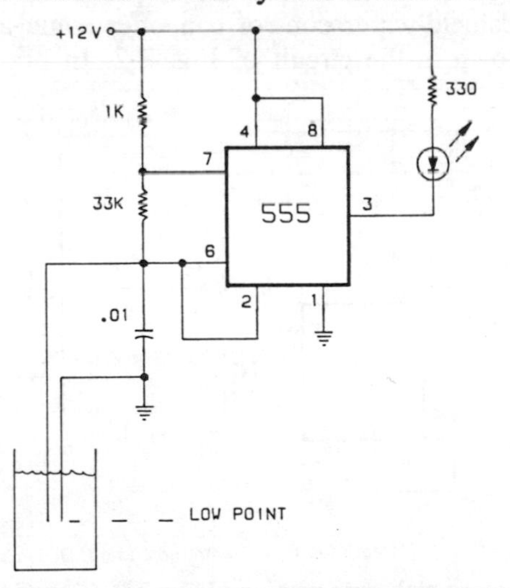

Fig. 7-1. Windshield-washer fluid monitor circuit.

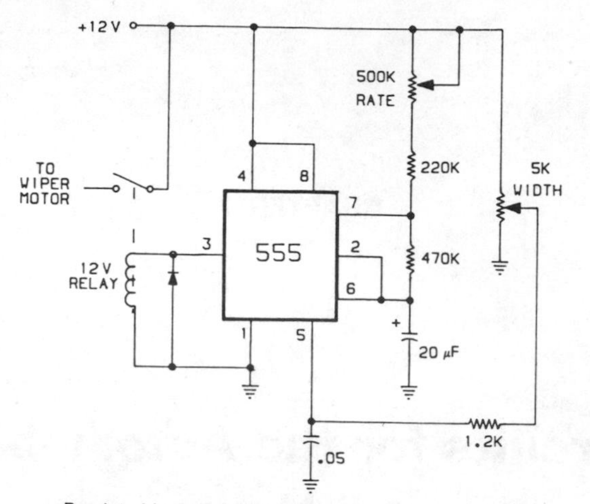

Reprinted by permission of *Popular Electronics* magazine.

Fig. 7-2. A windshield-wiper control that operates intermittently. (Copyright © 1975 Ziff-Davis Publishing Company.)

insulated wire probes, with about 5 mm of insulation stripped from the end to be placed in the liquid, are connected across the external timing capacitor C1. When the liquid is in contact with both tips, C1 is shorted out so that the timer is, in effect, turned OFF. When the liquid drops below the probes, the astable timer is free to run. It is coupled to an 8 Ω speaker. If only a visual indicator is desired, an LED is connected to the timer's output.

When light rain and dew become a problem, an intermittently operating windshield-wiper control can offer some advantages. Such a control is shown in the circuit of Fig. 7-2. In operation, the relay

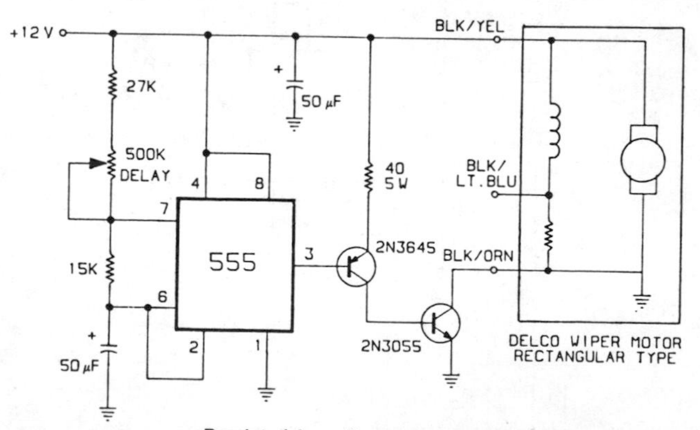

Reprinted from *Electronic Design,* December 20, 1974.

Fig. 7-3. A more elaborate version of Fig. 7-2. (Copyright © Hayden Publishing Company, Inc., 1974.)

is actuated at periodic intervals by the timer, closing the wiper motor contacts. The controls for the pulse rate and duration are easily adjusted for optimum performance after installation.

A more elaborate all solid-state version of the previous circuit is shown in Fig. 7-3 which allows the wiper to sweep at selected rates from once a second to once every 20 seconds.

AUTOMOBILE IDIOT-LIGHT ALARM

Ever wonder if the so-called "idiot lights" in the car's dash panel were burned out or loose when you are driving? These are the lights that are supposed to warn you that either the oil pressure is low, the engine temperature is too high, the generator is not charging, or the emergency brake has been left on. The simple alarm circuit of Fig. 7-4 is designed to warn you when any of these lights come on. If the idiot light does not light, for any reason, the individual LEDs will still indicate which light was supposed to be lit. Current to any of the lights will gate a 555 timer whose output drives a small speaker to also attract attention.

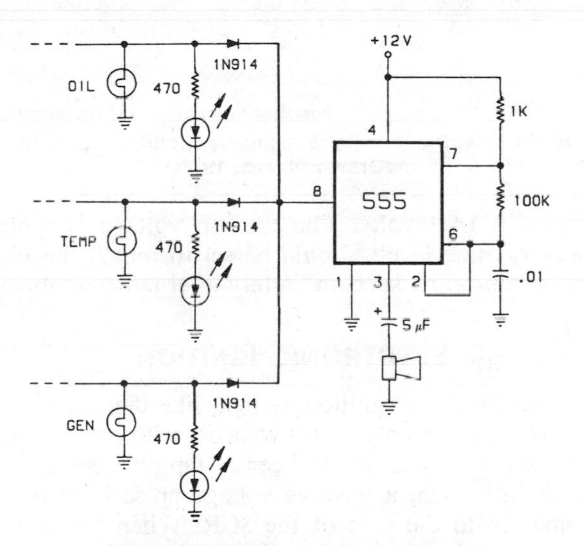

Fig. 7-4. A warning-light alarm circuit.

VOLTAGE REGULATOR

If you ever have to replace a faulty voltage regulator in your car, you might want to try the 555 timer regulator circuit shown in Fig. 7-5.

When the battery voltage becomes too low, the timer turns ON driving its output HIGH, and a current of 60 mA flows through R2.

This causes a sufficient bias voltage to be developed across R1, turning transistor Q1 on and energizing the alternator's field coil.

The regulator's low-voltage turn-on point is fixed by setting the timer's trigger voltage to approximately one half the reference voltage existing at its control pin, and which is established by the diode string D2–D5 (approximately 5.9 volts). The high-voltage turn-off point is set by making the timer's threshold voltage equal to the reference

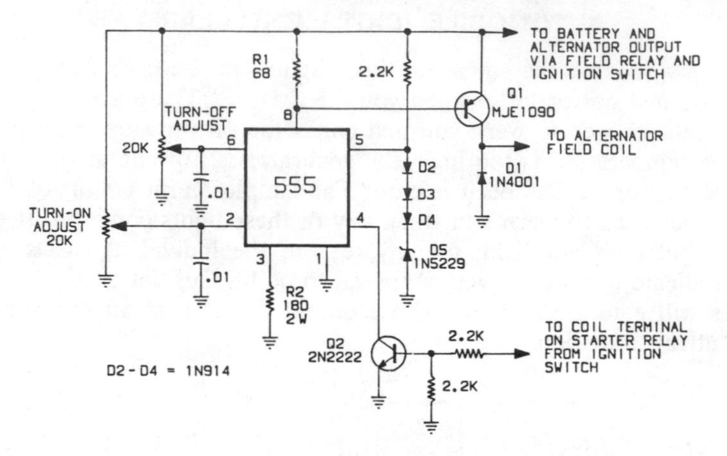

Reprinted from *Electronics*, February 21, 1974.

Fig. 7-5. Electronic voltage regulator circuit. (Copyright © McGraw-Hill, Inc., 1974.)

voltage, typically 14.9 volts. The turn-on voltage is typically 14.4 volts. However, these levels should be set to match the charging requirements of your car's specific battery-alternator combination.

ELECTRONIC IGNITION

A capacitive-discharge ignition system, like the one shown in Fig. 7-6, can deliver approximately 500 volts dc, with a 555 timer to drive the converter section using a 6.3-V center-tap filament transformer.

With the points open, a positive voltage pulse is coupled through R10, D8, and C4 to the gate of the SCR. When the SCR fires, C2 discharges through the spark coil and starts to recharge with opposite polarity. This recharge cycle provides a negative charge through R8 and D8 to the SCR, preventing its retriggering after the SCR turns OFF.

When the points close, C4 is discharged through R9 and R10 so that the SCR can be retriggered. The time required for this discharge provides a delay to prevent erratic SCR firing caused by "point bounce" at high engine rpm.

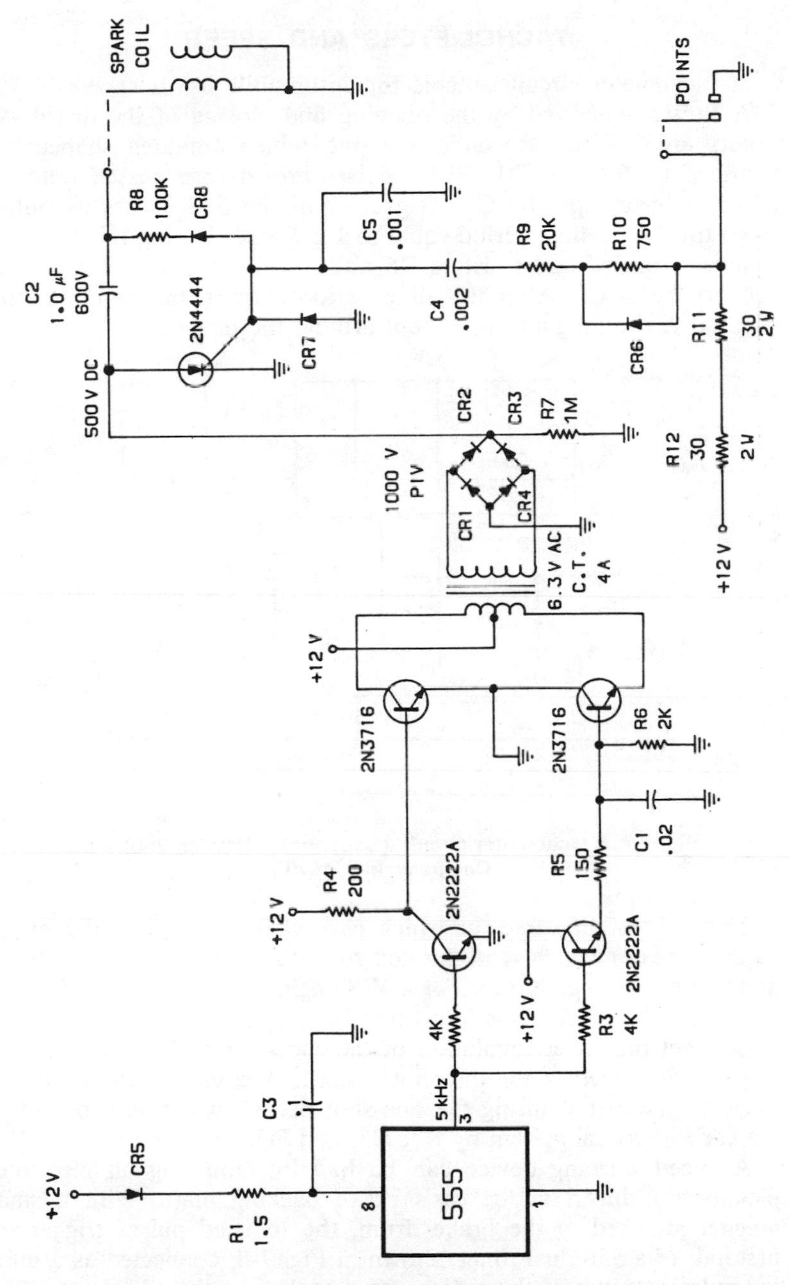

Reprinted from *Electronic Design,* November 22, 1974.

Fig. 7-6. A capacitive-discharge ignition system. (Copyright © Hayden Publishing Company, Inc., 1974.)

TACHOMETERS AND SPEED

A tachometer circuit suitable for automobile use is shown in Fig. 7-7. Pulses generated by the opening and closing of the distributor points are fed into the circuit's input, which are then shaped and clamped by R1 and D1. These pulses are in turn passed onto the timer's trigger input by C1. Triggering of the 555 causes its output to go HIGH for a time period equal to 1.1 R4C2. During this time, D2 is reverse biased and resistors R5 and R6 provide a calibrated current to the meter. After this time period elapses, the timer's output goes LOW, shunting all the current around the meter.

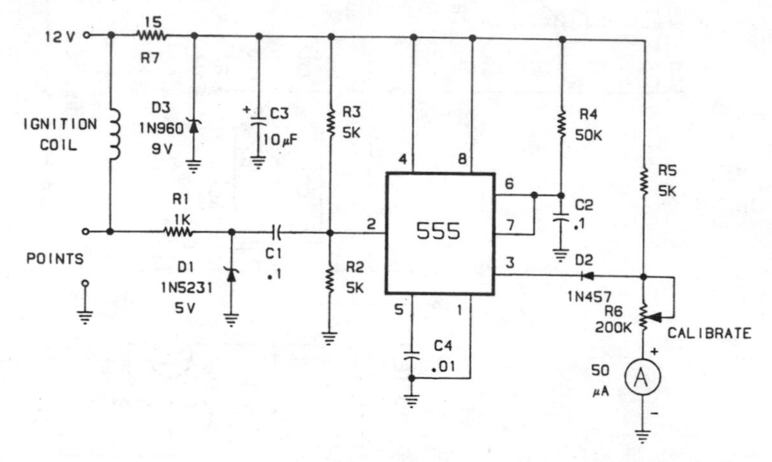

Reprinted from *Electronic Design*, June 7, 1973.

Fig. 7-7. A tachometer circuit. (Copyright © Hayden Publishing Company, Inc., 1973.)

The ratio of the time for which current flows through the meter, to the time for which it is shunted to ground, gives an accurate indication of the engine rpm. For a V-8 engine, the frequency of pulses at the distributor points is four times the engine rpm, since the points close eight times per revolution of the camshaft and the engine runs at twice the speed of the distributor shaft. A constant-current source is fed to the meter during the one-shot period, which is supplied by the car's electrical system by R7, C3, and D3.

A speed warning device can be had by mounting an electronic pickup transducer on the car's brake backing plate. With a small magnet attached to the brake drum, the induced pulses trigger the first half of a 556 dual timer, shown in Fig. 7-8, connected as a missing pulse detector. If the pulses occur each time the wheel rotates at a low enough frequency, the output from the first timer will alternate between its HIGH and LOW states. The second half of the 556, also connected as a missing pulse detector, is driven from the output of the first. If the pulses continue to arrive at the input of the second timer, its output remains HIGH. When the "speed" setting is exceeded, the pulses occur too rapidly for the time constant of the first timer to respond, causing its output to remain HIGH. After the delay, this in turn causes the output of the second timer to go LOW, activating the warning indicator.

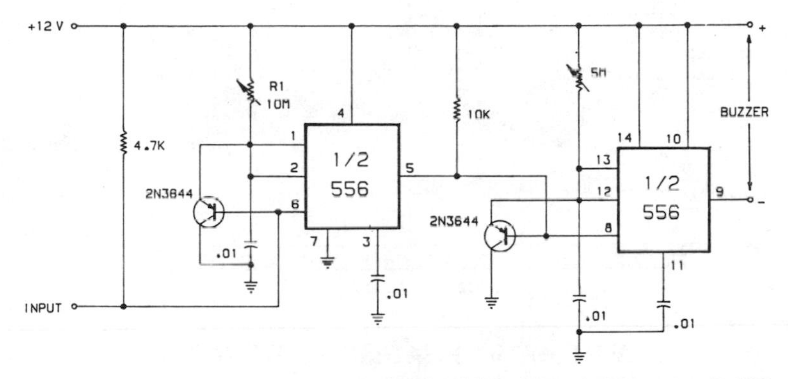

Reprinted from *Electronic Design*, June 7, 1973.

Fig. 7-8. A 556 dual timer connected as a dual missing pulse detector. (Copyright © Hayden Publishing Company, Inc., 1973.)

The necessary time constants can be calculated from the car's tire size. For example, a tire with an outer diameter of 25 inches covers a distance of 78.5 inches per revolution. Since 1 mph equals 1.467 feet per second, each revolution can then be equated to miles per hour. Therefore, R1 can be either a calibrated potentiometer or several fixed resistors and a rotary switch.

AUTOMOBILE BURGLAR ALARM

Another dual-timer application is the automobile burglar alarm circuit of Fig. 7-9. The first half of the 556 provides a convenient time delay for "arming" the system, and also allows the driver to enter the car and disarm the circuit. This feature eliminates the need for the inconvenient and often vulnerable arming switch on the outside of the car. Consequently, the OFF/ON switch can be placed out of sight underneath the car's dashboard.

When the alarm goes off, the second half of the timer is triggered by the first. After the initial "turn-on," the SCR prevents the second timer from triggering itself until one of the grounding-type sensor switches triggers the timer.

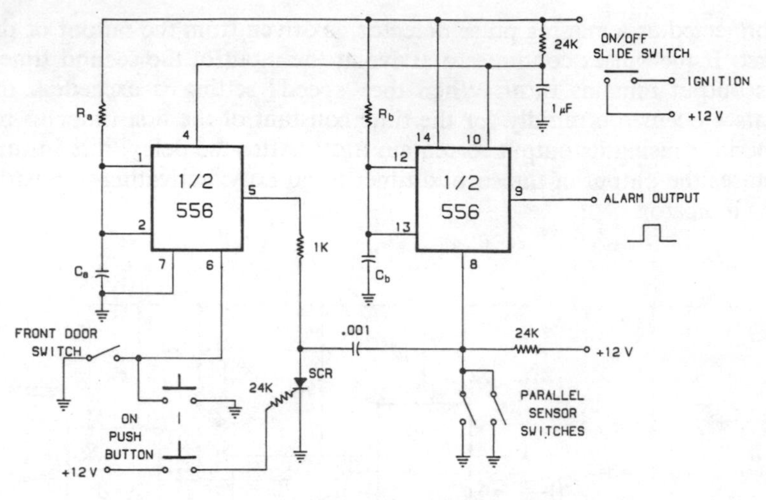

Reprinted from *Electronics*, June 21, 1973.

Fig. 7-9. An automotive burglar alarm circuit. (Copyright © McGraw-Hill, Inc., 1973.)

AUTOMATIC HEADLIGHT TURN-OFF

Anyone who has stumbled around in the dark, especially in a garage after leaving the car at night, will appreciate the automatic circuit of Fig. 7-10. It is used to automatically turn off the headlights at some convenient time after the ignition is switched OFF.

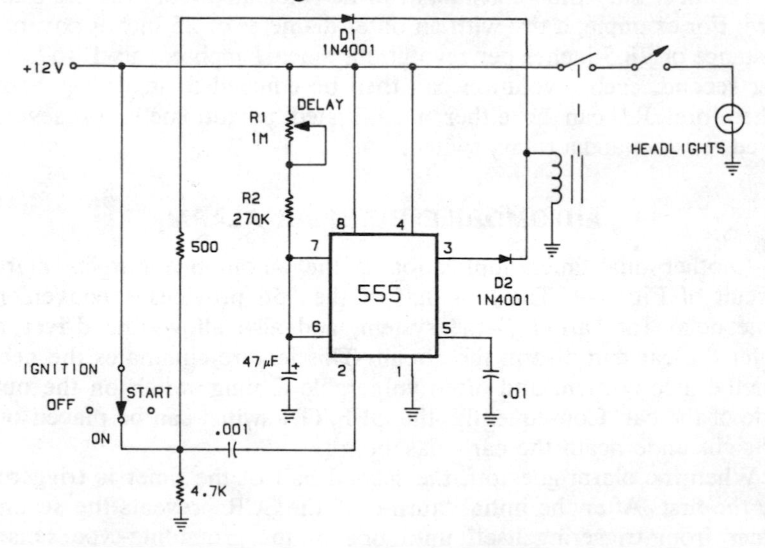

Fig. 7-10. Circuit for turning car headlights off automatically.

When the ignition is first switched ON, the battery voltage is fed to the relay coil through the 500 Ω resistor and diode D1. Switching OFF the ignition generates a negative-going pulse on pin 2 that triggers the timer. The timer's output then energizes the relay to keep the lights on long enough to leave the area. The delay is adjustable from approximately 10 to 60 seconds with the values shown in Fig. 7-10.

TOXIC GAS ALARM

There is always danger associated with the combustible gases near propane-fired camper stoves, gasoline fumes, and the exhaust released from engines, especially if the gas is allowed to accumulate in a confined area. The gas alarm, shown in Fig. 7-11, uses a tin oxide semiconductor sensor (available from Southwest Technical Products Co., 219 W. Rhapsody, San Antonio TX 78216) heated by a thin filament coil. The sensor lowers its resistance when exposed to a variety of gases such as hydrogen, carbon monoxide, propane, and alcohols, to name a few.

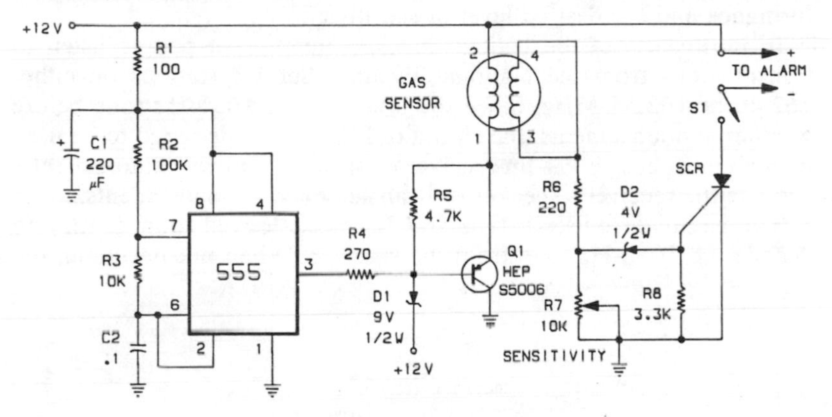

Reprinted by permission of *Elementary Electronics* magazine.

Fig. 7-11. Toxic gas alarm circuit. (Copyright © 1974 Davis Publishing, Inc.)

The alarm-tripping circuit turns the buzzer ON when the sensor's resistance decreases so that the voltage at the gate of the SCR exceeds a value preset by the sensitivity control R7. Once triggered, the buzzer sounds, and switch S1 must be used to reset the SCR. Zener diode D2 prevents the circuit from triggering if a transient appears on the +12-V power line.

The unique part of this circuit is the power supply built around the 555 timer. The timer sends out periodic pulses that turn Q1 ON and OFF, thus gating the battery voltage. This approach consequently saves approximately 80% of the battery power, as compared to the conventional voltage-dropping resistor or power-transistor heat-sink methods. Zener diode D1 provides a constant filament-supply voltage for the sensor.

Since this sensor element has a fair amount of thermal inertia, switch S1 should be in the OFF position for 4 to 5 minutes after the unit is turned on. Potentiometer R7 then should be adjusted for the desired sensitivity, as the sensor is easily capable of responding to concentrations of as little as 100 ppm of carbon monoxide.

MONITORING THE WEATHER

A simple storm-warning device, coupled with a transistor radio, can be built around a 555 timer, as shown in Fig. 7-12. The circuit depends upon the accumulation of a charge across C1 by static noise bursts received from an external am radio. When the charge on C1 builds up, both the SCR and the timer are triggered, with the SCR applying power to the buzzer and the LED for a visual indication. In operation, R1, R2 and R3 will have to be adjusted for optimum performance and the desired level of sensitivity.

In most parts of the United States, a number of people listen to transmissions from the National Weather Service stations on either 162.40 or 162.55 MHz. Since the NWS uses a 1050-Hz tone before weather announcements, the circuit of Fig. 7-13 is designed to be used with an inexpensive vhf-fm receiver or scanner and will automatically activate the receiver's speaker to hear the voice announcements.

The circuit first uses a type-567 IC phase-locked loop (PLL) to decode the 1050-Hz tone from the receiver. When not decoding, the

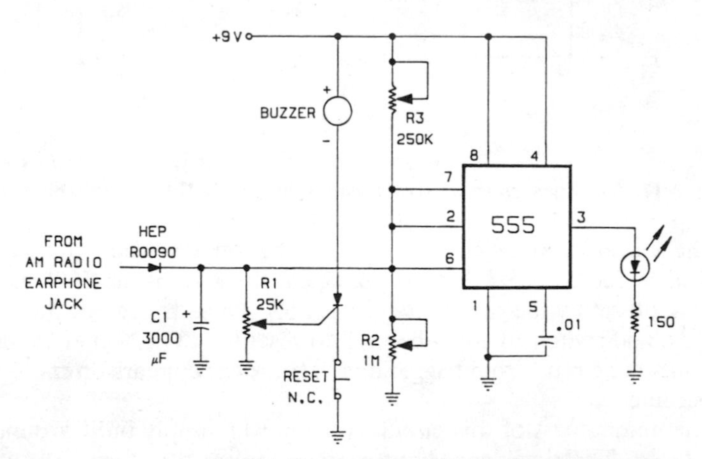

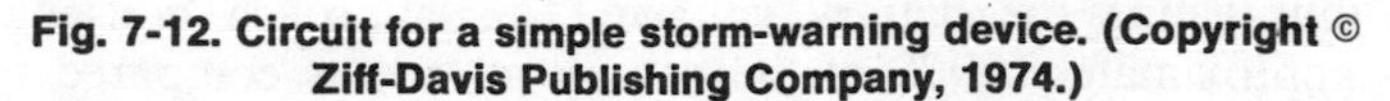
Reprinted by permission of *Popular Electronics* magazine.

Fig. 7-12. Circuit for a simple storm-warning device. (Copyright © Ziff-Davis Publishing Company, 1974.)

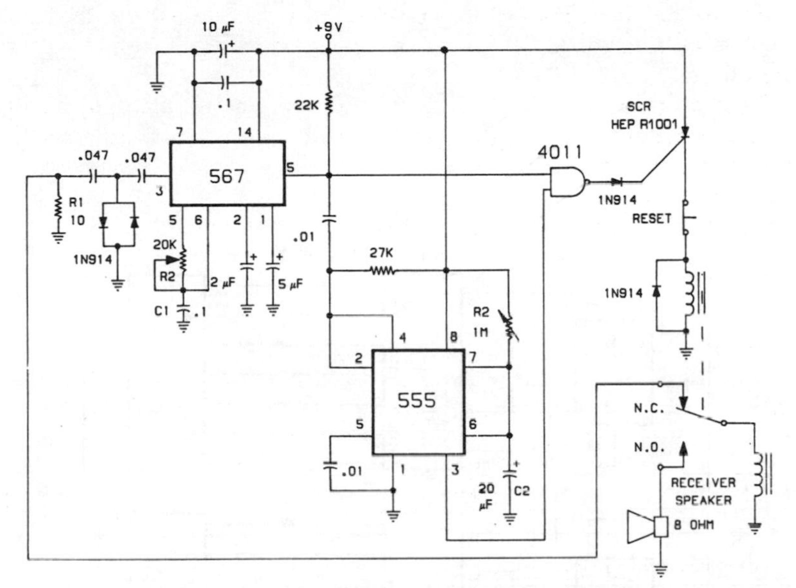

Reprinted by permission of *Popular Electronics* magazine.

Fig. 7-13. An automatic monitoring circuit. (Copyright © 1976 Ziff-Davis Publishing Company.)

PLL output, at pin 8, is HIGH. When the PLL receives a tone within its locking range, the output of the PLL goes LOW, triggering the 555 timer. The timer is required because false alarms often are produced by random receiver noise or the voice announcements that often are within the 1050-Hz range. Since these "signals" are usually of short duration while the 1050-Hz alert tone is transmitted for at least 15 seconds, the timer then acts as a 10-second noise filter.

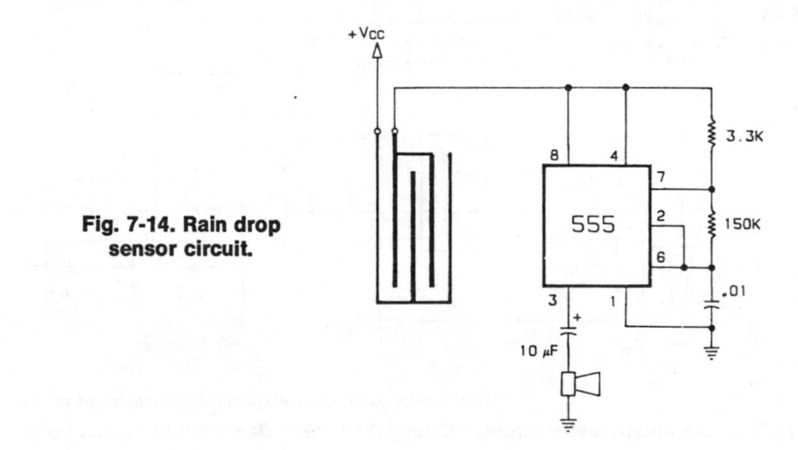

Fig. 7-14. Rain drop sensor circuit.

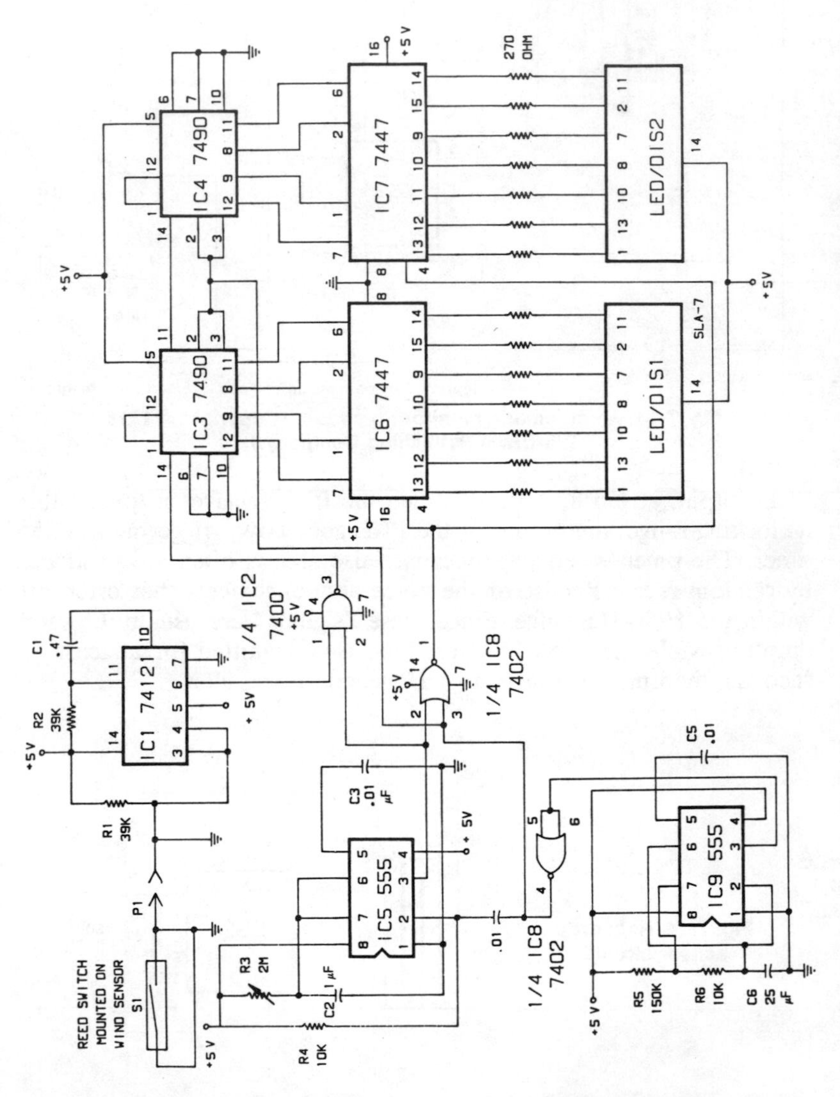

Reprinted by permission of *Electronics Hobbyist* magazine.

Fig. 7-15. An anemometer circuit. (Copyright © 1976 Davis Publications, Inc.)

As long as the alert tone is not present, the SCR does not conduct, and the relay is not energized. However, when the 1050-Hz tone is received, the PLL locks and the timer starts its cycle. At the end of 10 seconds (determined by R3 and C2), the SCR conducts and the receiver's speaker is connected permitting the voice announcement to be heard. Otherwise, the receiver's speaker is normally disconnected. Thus, the receiver is quiet, with resistor R1 providing a substitute load.

With a 1050-Hz audio-signal generator placed across R1, potentiometer R2 is adjusted until the PLL's output goes LOW. Then R3 is adjusted to give about a 10–15 second delay. After the alarm sounds and the announcement is heard, the circuit is simply reset by pressing S1.

A simple circuit that is an extension of the circuit of Fig. 5-3 indicates the presence of falling rain drops. It uses a grid of closely spaced parallel contacts as shown in Fig. 7-14. When any rain drops hit the sensor, the contacts are shorted and the timer is turned ON, sounding the alarm.

Another circuit that uses the tachometer approach is an anemometer to measure the wind velocity. In Fig. 7-15, the wind turns a shaft with a reed switch, actuated by rotating magnets. Each time the cups make a full revolution, the switch opens and closes twice, triggering a 74121 TTL one-shot to "clean up" the pulses. These pulses are then fed to a NAND gate controlled by a 555 one-shot (IC5), whose output pulse width is adjusted for calibration. Another 555 timer (IC9) provides automatic triggering pulses for IC5, as well as supplying reset and blanking pulses for the digital counters and decoder/drivers.

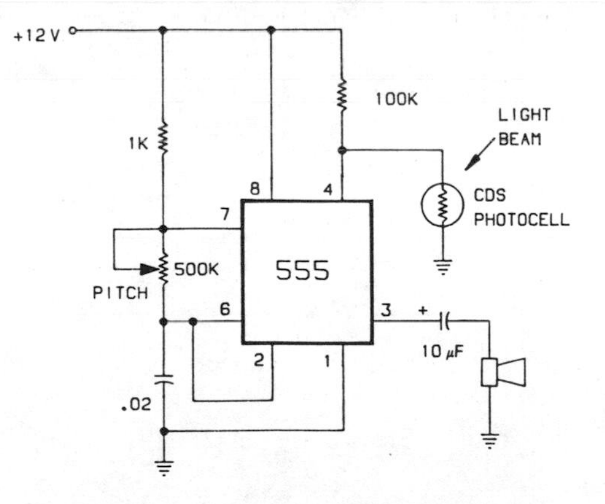

Reprinted by permission of *Popular Electronics* magazine.

Fig. 7-16. Electric-eye alarm circuit. (Copyright © 1974 Ziff-Davis Publishing Company.)

ELECTRIC EYES

An extension of the timer circuits using photocells, in Chapter 5, is the electric-eye doorway or window alarm shown in Fig. 7-16. When there is light striking the photocell, the timer's reset pin is held LOW, disabling the timer. When the light beam is interrupted, the reset pin is forced HIGH, and the timer is free to oscillate. This circuit can then be coupled to a 555 one-shot multivibrator to function as a photoelectric counter driving a 7-segment digital display.

CHAPTER 8

The 555 and Ma Bell

This chapter contains circuits and technical information which is public knowledge and has been previously published in many magazines and journals. However, since all of these circuits involve the use of telephone lines and equipment to some degree, they nevertheless are presented only to illustrate the unique uses of the 555 timer. Consequently, the phone companies do not hesitate to prosecute anyone tampering with their phone lines.

RINGERS

The ringing pulse signal on a telephone line can be made to operate a remote ringer without overloading the phone lines, without interfering with company service, and without degrading operation of the line receiving the signal, if it is passed through an optocoupler. In Fig. 8-1, the optocoupler transfers the ringing signal to the rest of the cir-

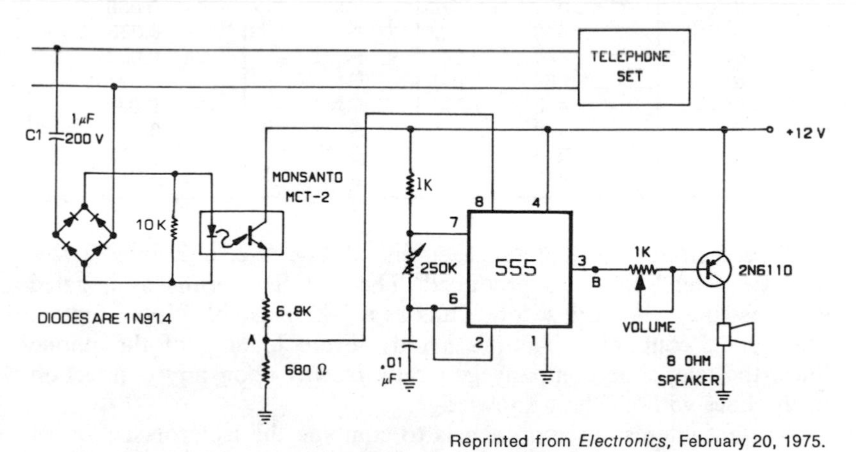

Reprinted from *Electronics*, February 20, 1975.

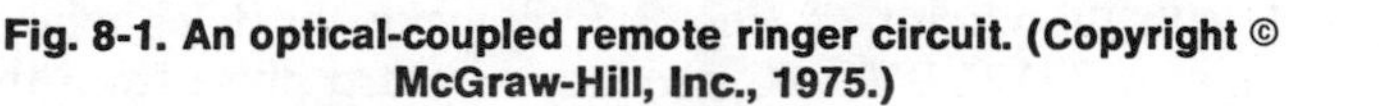

Fig. 8-1. An optical-coupled remote ringer circuit. (Copyright © McGraw-Hill, Inc., 1975.)

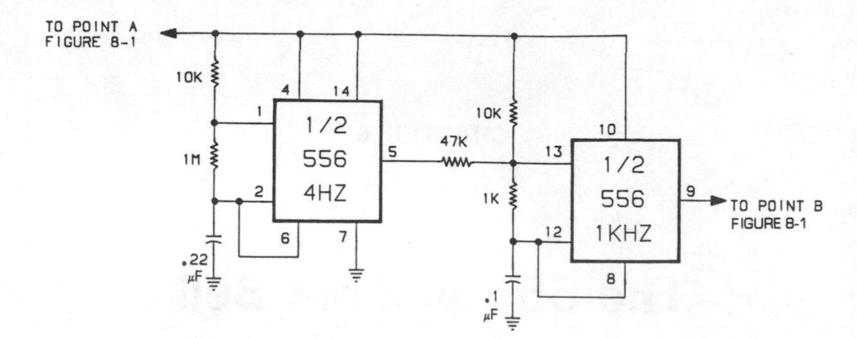

Fig. 8-2. Circuit modification to produce a 2-note sound.

cuit and also isolates this circuitry from the telephone line. The output current from the optocoupler triggers a 555 timer whose output drives a remote speaker that sounds whenever a ringing signal is present on the lines.

A telephone ringing signal of about 100 volts at 20 Hz has a cycle of 2 seconds ON and 4 seconds OFF. The signal is applied to the octocoupler through C1 and the diode bridge which doubles the frequency to 40 Hz. This is done because a 40-Hz gating rate sound from the speaker is more pleasing to the ear than a 20-Hz rate.

Table 8-1. Tone Frequencies for Fig. 8-3

Count	Frequency (Hz)	Note	Capacitor (µF)
0	329	E	0.042
1	349	F	0.040
2	370	F#	0.038
3	440	A	0.033
4	370	F#	0.038
5	392	G	0.036
6	415	G#	0.034
7	440	A	0.033
8	523	C	0.027
9	440	A	0.033

By using the 556 dual timer circuit of Fig. 8-2, a 2-note "twee-dell, twee-dell" sound is produced. The first timer runs as a gated 4-Hz astable, while the second runs as a 1-kHz astable. Even though this optical-coupling technique avoids severe loading of the phone lines, the telephone company generally frowns upon any connection to the lines without their knowledge.

Another interesting approach is to combine the isolator section of Fig. 8-1 with the circuit of Fig. 8-3 to make the telephone play a simple tune. A 74121 TTL IC produces a sharp pulse from the tele-

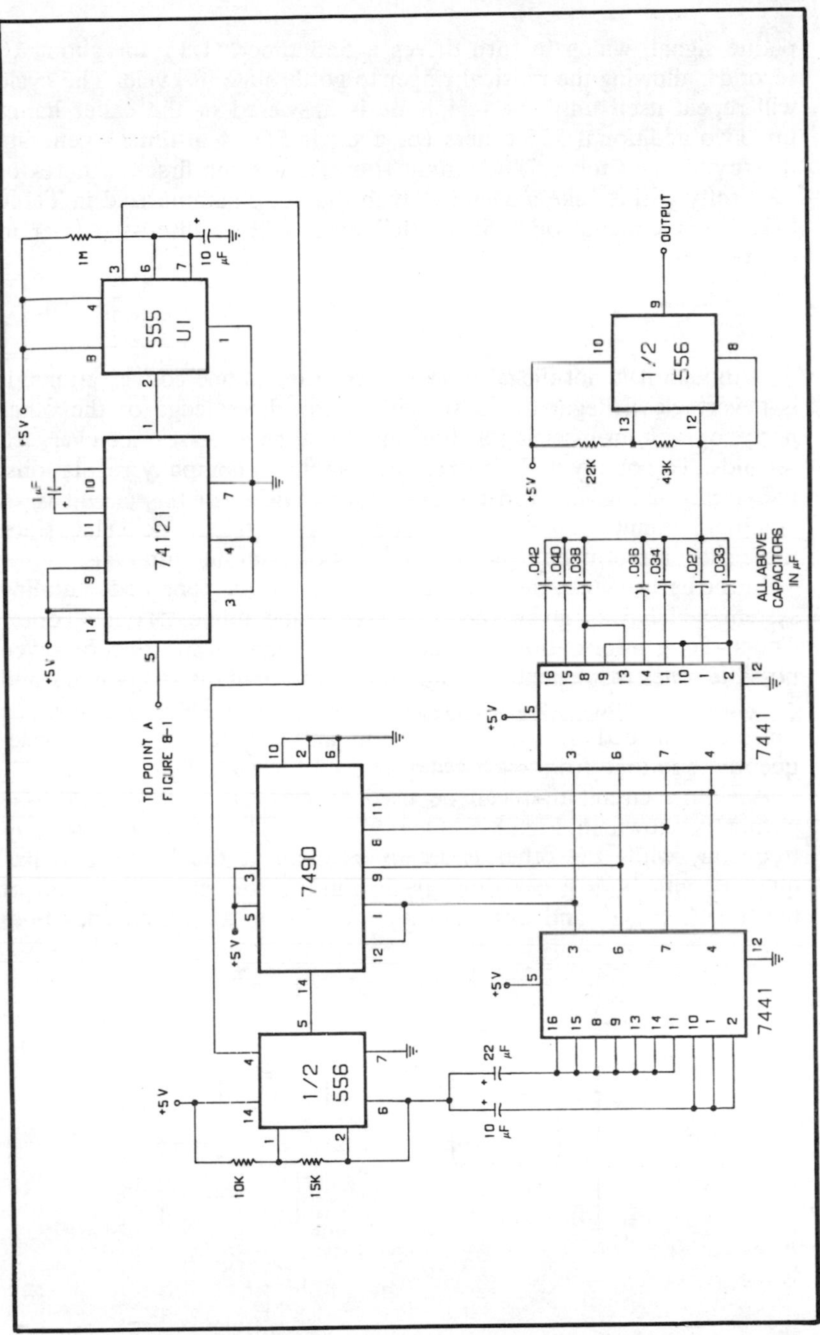

Reprinted from *Electronics*, May 15, 1975.

Fig. 8-3. A musical-tone ringer circuit. (Copyright ©McGraw-Hill, Inc., 1975.)

phone signal, which in turn drives a 555 timer (U1) for about 10 seconds, allowing the musical ringer to go through its cycle. The cycle will repeat itself until the telephone is answered or the caller hangs up. Two additional 555 timers (or a single 556 dual timer) generate the rhythm and tones. The circuit shown plays the first ten notes of "A Pretty Girl is Like a Melody" with the notes summarized in Table 8-1. Other circuits for "555 music" tunes will be discussed later in Chapter 9.

TONES

Although it is not illegal to tape record telephone conversations, it is nevertheless illegal to do so without the knowledge of the other party, or without inserting a short audible tone at least once every 15 seconds. To comply with federal and telephone company regulations, the circuit of Fig. 8-4 can be used. An extension of this circuit is an audible 1-minute reminder for long distance telephone calls, since these rates are now charged on the basis of 1-minute intervals.

In the early 1960s, the Bell System introduced a tone-coded dialing system which was given the registered trade name "Touch-Tone." Touch-Tone information is coded in tone-pairs, using two of seven possible tones for digits 0 through 9, and the symbols # (pound) and * (star). The audio frequencies used are given in Table 8-2, and the tones are divided into a "low" group and a "high" group. Consequently, one tone from each group is used for each digit.

A simple circuit that can be used to generate a single two-tone output is shown in Fig. 8-5. One 555 timer is set to a low-group frequency while the other is set to oscillate at the high-group frequency, with both generating "triangular" waves at the junction of the timer's trigger and threshold inputs. The op amp then adds both

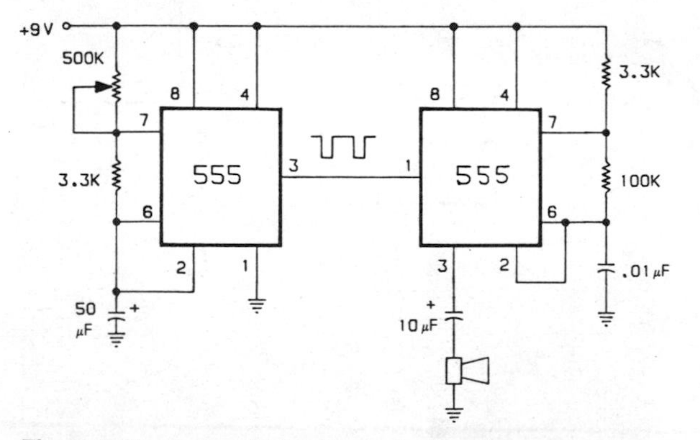

Fig. 8-4. Tone-signal circuit used when tape recording telephone conversations.

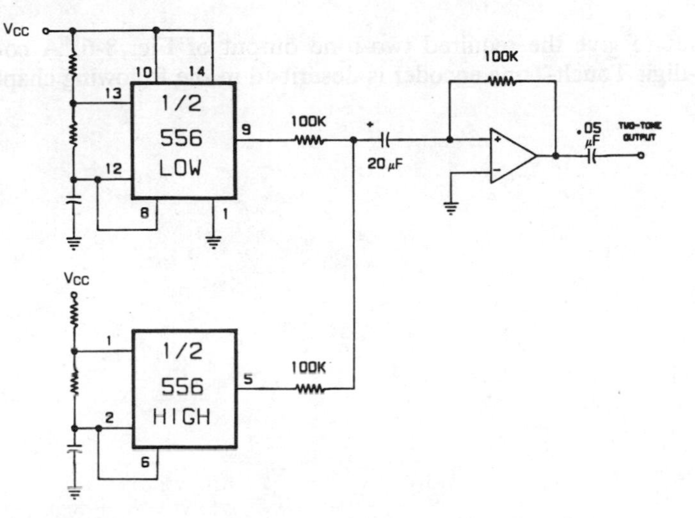

Fig. 8-5. Simple circuit for generating a two-tone output.

Table 8-2. Audio Frequencies Used for Touch-Tone® Coding

Low Tone Group (Hz)	High Tone Group		
	1209 Hz	1336 Hz	1477 Hz
697	1	2	3
770	4	5	6
852	7	8	9
941	*	0	#

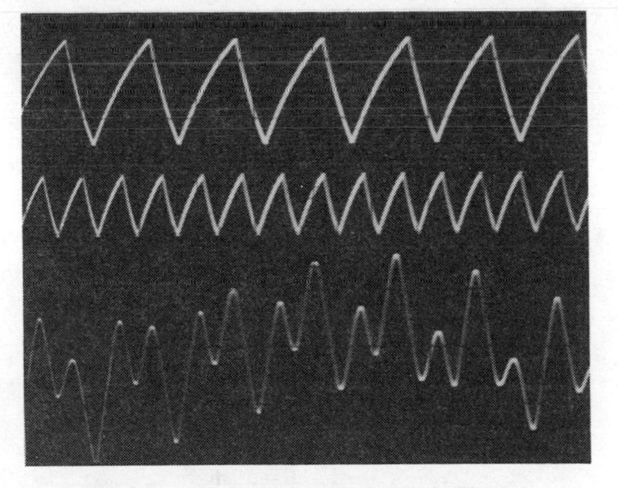

Fig. 8-6. Output of op amp shown in Fig. 8-5.

tones to give the required two-tone output of Fig. 8-6. A complete 12-digit Touch-Tone encoder is described in the following chapter.

CHAPTER 9

Hobbies

This chapter describes the applications of the 555 timer for use in photography, music, Citizens Band and amateur radio.

PHOTOGRAPHY

In most cases, the development time for Polaroid pictures depends on the type of film used. Although commercial timers are available, the circuit of Fig. 9-1 will give an audible alarm at the end of the selected development period. When the switch is pressed, the first timer is triggered causing its output to go HIGH for the selected period set by the external timing resistors. At the end of the desired period, the output then goes LOW, triggering the next timer which is connected as a 1-second one-shot and, thus, gating the audible alarm. For simplicity, an ordinary 8Ω earphone from a transistor radio can be used instead of a speaker. The 15-second time period is used for type 51,

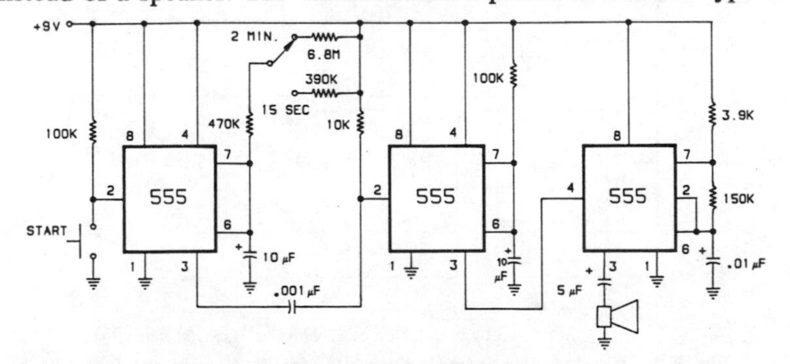

Fig. 9-1. Audible alarm circuit for Polaroid film development times.

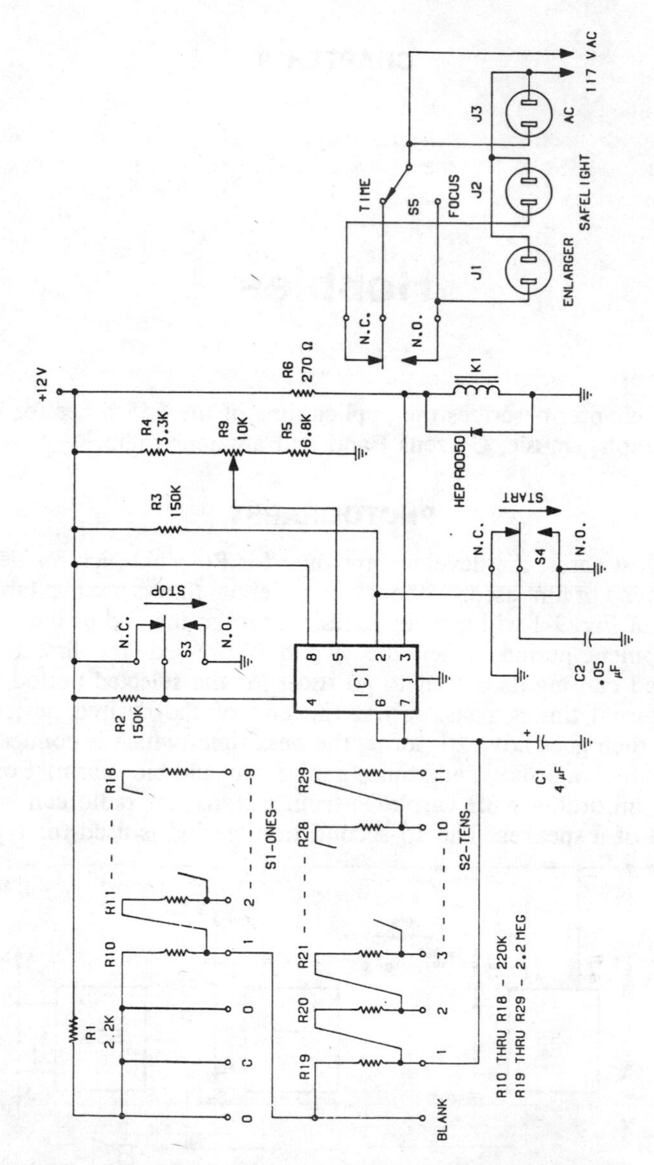

Reprinted by permission of *Electronics Hobbyist* magazine.

Fig. 9-2. An enlarger printmeter control circuit. (Copyright © 1976 Davis Publications, Inc.)

52 and 57 Polaroid films, while the 2-minute position is used for type 46 film.

For those who prefer to do their own darkroom developing, the circuit of Fig. 9-2 can be used to control an enlarger printmeter. A 555 timer is connected as a monostable with automatic reset. The timing interval is therefore set by C1 and the resistors which are selected by switches S1 and S2. With the values shown, the timing is adjustable in 1-second steps from 1 to 119 seconds. Potentiometer R9 is used to vary the timing control voltage to allow for variation in the tolerance of C1 (see Chapter 2). The enlarger's power cord plugs into socket J1, the safelight into J2, and the enlarger exposure meter into J3. When either focusing or using the exposure meter, switch S5 is set to FOCUS. Consequently the safelight will be OFF. For print exposures, S5 is switched to TIME and S4 is pressed. During the exposure, the safelight will be off, but will automatically turn ON upon completion of the exposure.

On the other hand, by using a photocell with a 555 timer as shown in Figure 9-3, we have a device that will automatically select the correct exposure and expose the print for the correct amount of time.

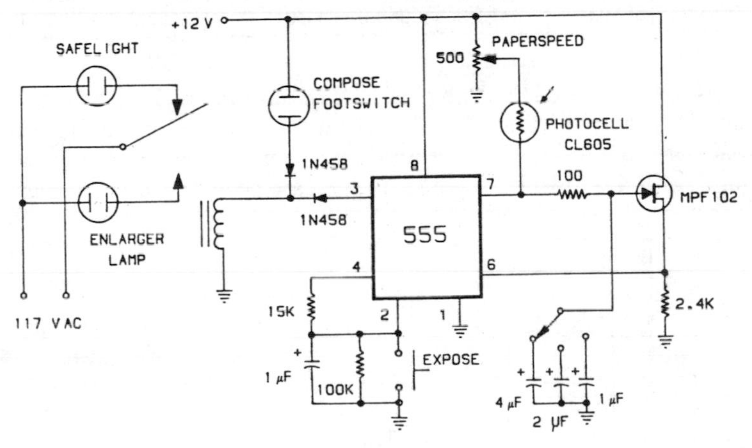

Reprinted by permission of *Popular Electronics* magazine.

Fig. 9-3. An automatic enlarging exposure circuit. (Copyright © 1974 Ziff-Davis Publishing Company.)

MUSIC

In Chapter 8, we briefly discussed how the 555 timer can be used to play music. The musical scale is divided into octaves, with each octave having 12 notes: C, C#, D, D#, E, F, F#, G, G#, A, A#, and B. For the normally used equally tempered scale, each note is $2^{1/12}$ higher than its immediate neighbor, resulting in a frequency ratio of

Table 9-1. Standard Frequencies (Hz) for the 12-Note Equally Tempered Musical Scale

Octave #	C	C#	D	D#	E	F
0	16.352	17.324	18.354	19.445	20.602	21.827
1	32.703	34.648	36.708	38.891	41.203	43.654
2	65.406	69.296	73.416	77.782	82.407	87.307
3	130.81	138.59	146.83	155.56	164.81	174.61
4	261.63	277.18	293.66	311.13	329.63	349.23
5	523.25	554.37	587.33	622.25	659.26	698.46
6	1046.5	1108.7	1174.7	1244.5	1318.5	1396.9
7	2093.0	2217.5	2349.3	2489.0	2637.0	2793.8
8	4186.0	4434.9	4698.6	4978.0	5274.0	5587.7
Octave #	**F#**	**G**	**G#**	**A**	**A#**	**B**
0	23.125	24.500	25.957	27.500	29.135	30.868
1	46.249	48.999	51.913	55.000	58.270	61.735
2	92.499	97.999	103.83	110.00	116.54	123.47
3	185.00	196.00	207.65	220.00	233.08	246.94
4	369.99	392.00	415.30	440.00	466.16	493.88
5	739.99	783.99	830.61	880.00	932.33	987.77
6	1480.0	1568.0	1661.2	1760.0	1864.7	1975.5
7	2960.0	3136.0	3322.4	3520.0	3729.3	3951.1
8	5919.9	6271.9	6644.9	7040.0	7458.6	7902.1

1.05946:1, as listed in Table 9-1. The circuit of Fig. 8-3 can be changed to play the first 10 notes of any tune by selecting the proper capacitors to produce the frequencies given in Table 9-1. In addition, Fig. 9-4 shows how a 555 timer can be wired as a trigger for

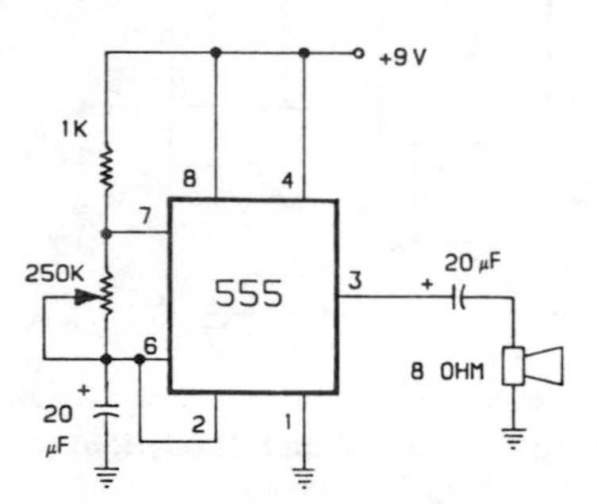

Fig. 9-4. Simple music rhythm or practice metronome circuit using a 555 timer.

electronic music rhythm or as a simple practice metronome when coupled to an 8Ω speaker.

HAM AND CB RADIO

Whether you are a beginning "ham," or have held a license for some time, a very handy item is the code practice oscillator. Using

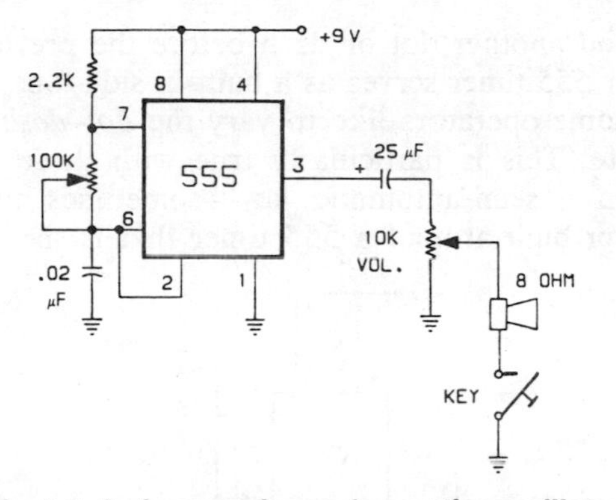

Fig. 9-5. An inexpensive code practice oscillator.

the 555 timer circuit of Fig. 9-5, one can have an inexpensive unit complete with pitch and volume controls.

A cw monitor to accompany your "on the air" sending is also helpful. In the circuit of Fig. 9-6, the rf signal from the transmitter is rectified by D1 and applied to the timer's reset input. When a positive pulse (key down) appears at the reset pin, the timer turns on.

Pretty soon, the beginning operator finds that there is a limit to how fast one can send with a "straight" key. In order to send faster with the proper spacing and ratio between the dots and dashes, an electronic keyer is often used. As shown in Fig. 9-7, the timer is the heart of this keyer. Operating as an astable clock, the sending speed is adjusted by the 50 kΩ potentiometer so that the maximum speed will be approximately 30 wpm. The proper 3:1 *dash–dot* ratio is formed by the two J-K flip-flops over the entire speed range. In addition, the characters are self-completing, which means that it is im

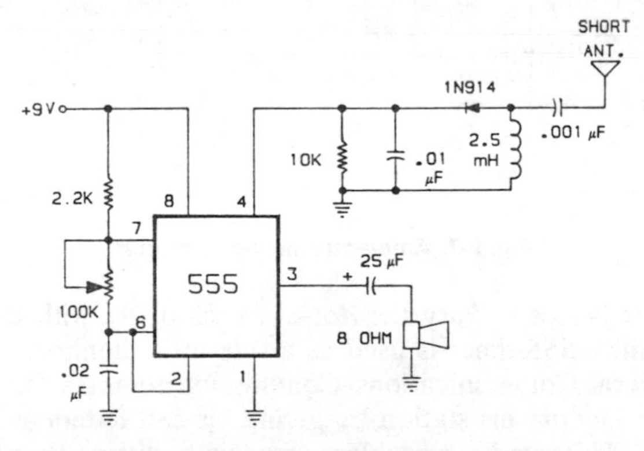

Reprinted by permission of *Popular Electronics* magazine.

Fig. 9-6. A cw monitoring circuit. (Copyright © Ziff-Davis Publishing Company.)

possible to send another dot or dash before the previous one is finished. Another 555 timer serves as a built-in side-tone monitor.

However, some operators like to vary the *dot–dash* length to suit their own taste. This is particularly true with those operators that originally used a semi-automatic key—sometimes called a "bug." Using the keyer built around a 555 timer that is shown in Fig. 9-8,

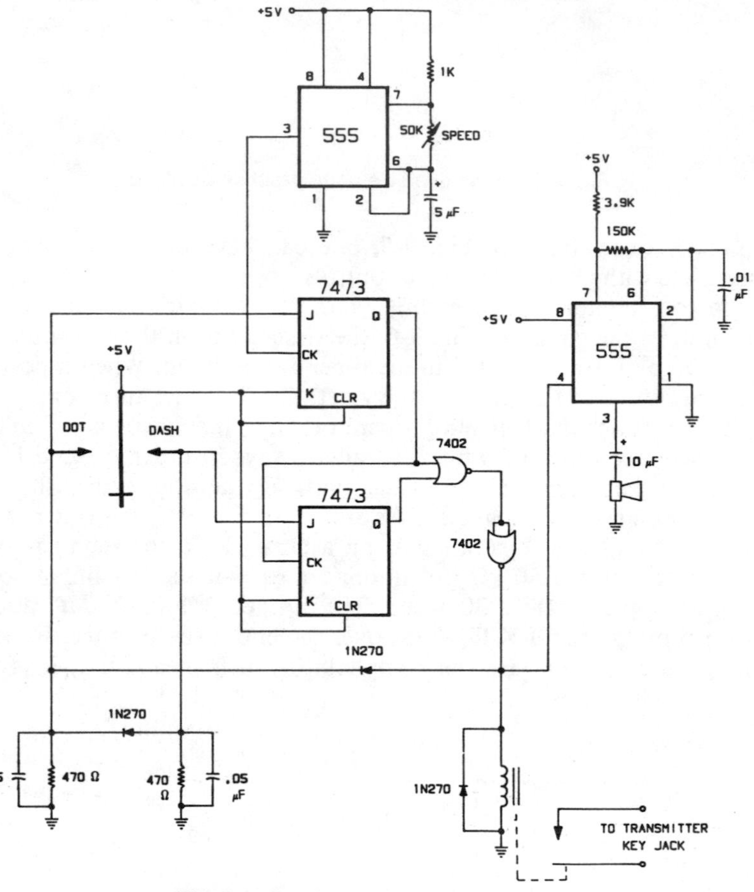

Fig. 9-7. An electronic keyer circuit.

the operator is free to vary the *dot–dash* ratio. As with the previous keyer circuit, a 555 timer is used as a side-tone monitor.

The Federal Communications Commission requires the ham radio operator to identify his station by giving his call letters at least every 10 minutes. This can be a problem, especially during lengthy conversations when it is difficult to keep track of the time. Compliance with this regulation is simplified by using a 555 timer as a one-shot as

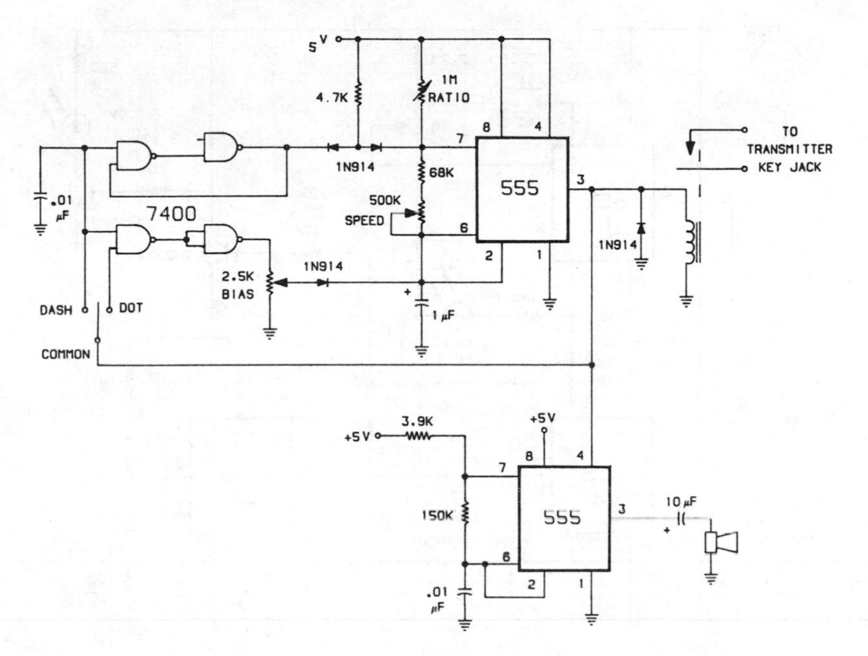

Reprinted by permission of *Ham Radio* magazine.

Fig. 9-8. A keyer circuit that can be adjusted by the operator. (Copyright © 1973 Communications Technology, Inc.)

shown in Fig. 9-9, so that a visual 10-minute warning indicator can be made. At the beginning of a conversation, the reset switch is pressed causing the green timing LED to light. After 10 minutes, as set by R1, the red LED will light to warn the operator that he must identify. The cycle can be reset at any time by simply pressing the reset switch.

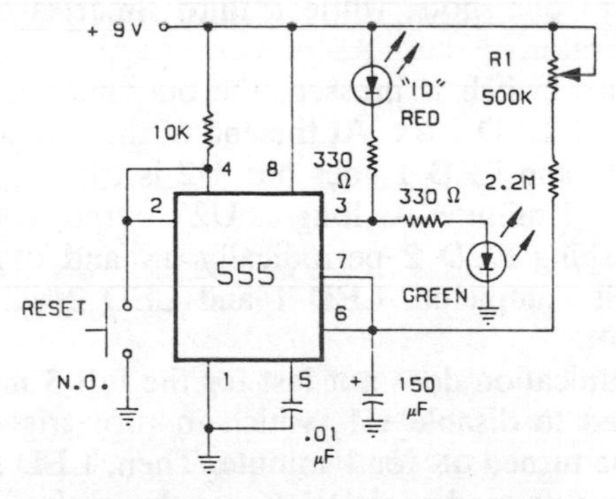

Reprinted by permission of *Ham Radio* magazine.

Fig. 9-9. 10-minute timer. (Copyright © 1974 Communications Technology, Inc.)

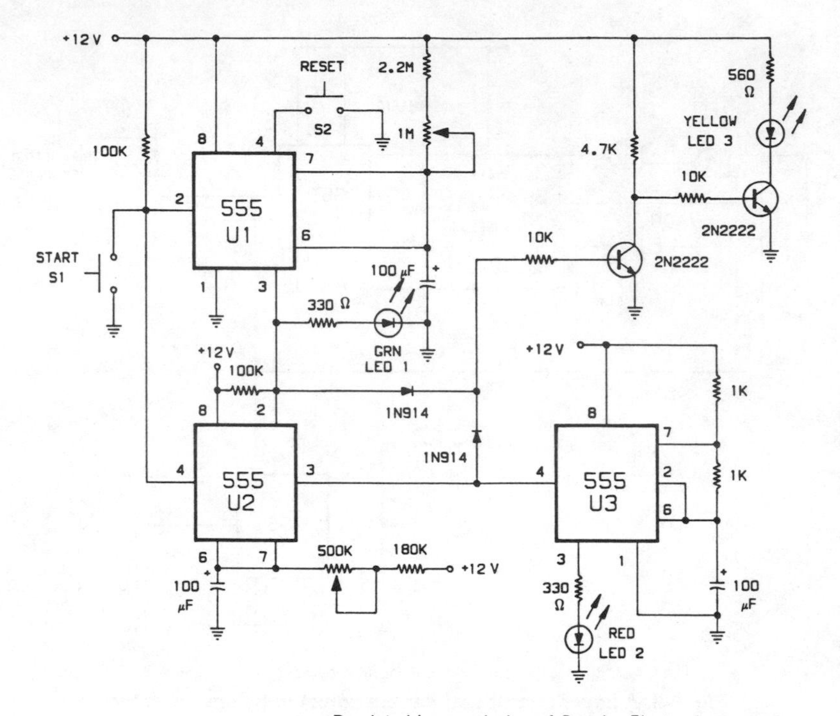

Reprinted by permission of *Popular Electronics* magazine.

Fig. 9-10. A timing circuit for CB operators. (Copyright © 1976 Ziff-Davis Publishing Company.)

However, for CB operators, the FCC requirements limit transmissions to a 5-minute "on" period followed by a 1-minute "off" period. The handy circuit of Fig. 9-10 will enable the fixed station or mobile operator to comply with the FCC rules. Two 555 timers, U1 and U2, are connected as one-shots, while a third timer is connected as an astable multivibrator.

When the start switch is pressed, the output of U1 remains HIGH for 5 minutes with LED 1 ON. At the end of the 5 minutes, the output of U1 goes LOW, and LED 1 goes out. U2 is triggered and its output remains HIGH for 1 minute. As long as U2's output is HIGH, U3 is free to oscillate, flashing LED 2 periodically ON and OFF. When the 1-minute period is completed, LED 1 and LED 2 will be OFF, while LED 3 comes ON.

If the communication does not last for the full 5 minutes, the reset switch is pressed to disable U1, which in turn triggers U2 and the oscillator will be turned ON for 1 minute. Then, LED 3 will light until the start switch is pressed again to repeat the cycle.

For amateur fm repeaters, the repeater's call letters must be sent at least once every 5 minutes. The circuit of Fig. 9-11 uses the 555

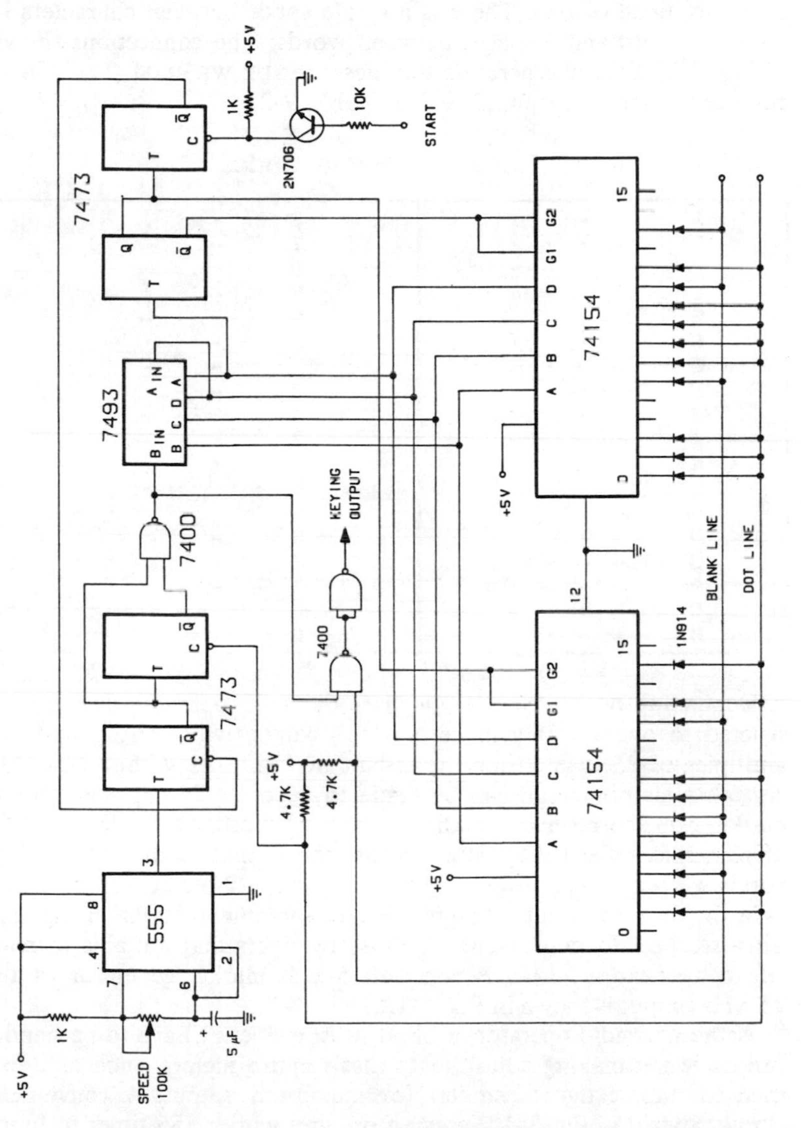

Reprinted by permission of 73 magazine.

Fig. 9-11. A Morse code call-sign generator. (Copyright © 1973, 73, Inc.)

timer to generate the call letters in Morse code. Using a pair of 4-line to 16-line decoders, a simple 32-position diode matrix is created for generating the call sign. For each dot, a diode is connected between the corresponding decoder output and the "dot" line. For each space, a diode is connected between the decoder and the "space" line. For a dash, no diode is used. There is a single space between characters in the same word and 3 spaces between words. The connections shown in Fig. 9-11 are for generating the message: DE WR3HAM. The Morse code characters are summarized in Table 9-2.

Table 9-2. Morse Code

A	· —	S	· · ·
B	— · · ·	T	—
C	— · — ·	U	· · —
D	— · ·	V	· · · —
E	·	W	· — —
F	· · — ·	X	— · · —
G	— — ·	Y	— · — —
H	· · · ·	Z	— — · ·
I	· ·	1	· — — — —
J	· — — —	2	· · — — —
K	— · —	3	· · · — —
L	· — · ·	4	· · · · —
M	— —	5	· · · · ·
N	— ·	6	— · · · ·
O	— — —	7	— — · · ·
P	· — — ·	8	— — — · ·
Q	— — · —	9	— — — — ·
R	· — ·	0	— — — — —

Depending on his class of license, the ham operator may be restricted to operate only in certain sub-bands, which are usually in multiples of 25 kHz. Using the circuit of Fig. 3-11 with a 100 kHz crystal, a simple circuit can be made to generate 25-kHz calibration markers on the receiver's audio. A pair of flip-flops divide the 100-kHz oscillator's signal by four, so that the output of the second flip-flop is a 25-kHz square wave. If desired, a type 7490 IC can be wired as a divide-by-5 counter to give 5-kHz incremental marker signals. This is especially convenient for those receivers that are able to read out to the nearest 5 kHz. A complete 5-kHz marker generator with a 25-kHz output is shown in Fig. 9-12.

As a ham radio operator, a blind person doesn't have to be handicapped when making adjustments that require meters, such as those used for tuning the transmitter for maximum output. A convenient circuit, shown in Fig. 9-13, uses an op amp with a 555 timer to function as a voltage-to-frequency converter. Consequently, the voltage developed across the meter is used to vary the frequency of the 555 astable section. If an ammeter is used, the voltage across it will be equal to its internal resistance, R_M, times the meter current, I_M, so that $V_M = I_M R_M$. The meter voltage is then amplified by an op amp having a differential gain of 100 (40 dB), which then drives the modulation input of the 555 timer. Consequently, the resultant pitch of the oscillator will be inversely proportional to V_M. Since blind persons have a remarkable sense of hearing, they are easily able to differentiate between small changes in pitch.

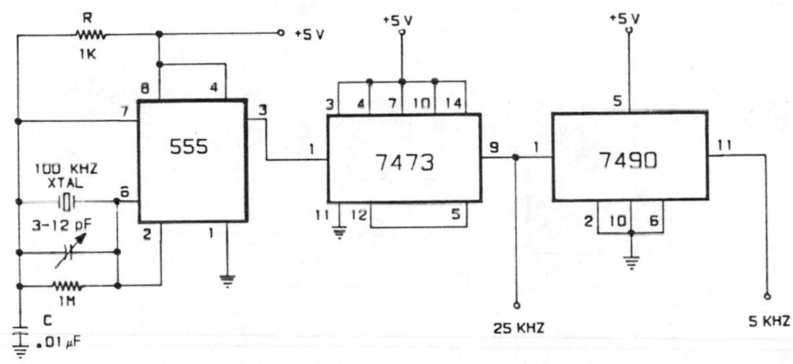

Fig. 9-12. A 5-kHz marker generator.

In Chapter 8, we discussed a circuit for a simple 1-digit Touch-Tone® encoder. However, for those using the "Autopatch" facilities of the local fm repeater, an encoder capable of generating all 12 digits of Table 8-2 must be used. As shown in Fig. 9-14, two 555 oscillators are used as before. However, the external timing resistor, normally placed between the timer's discharge pin and V_{cc} (R_a in Fig. 3-1A), is now replaced by a resistor-divider string.

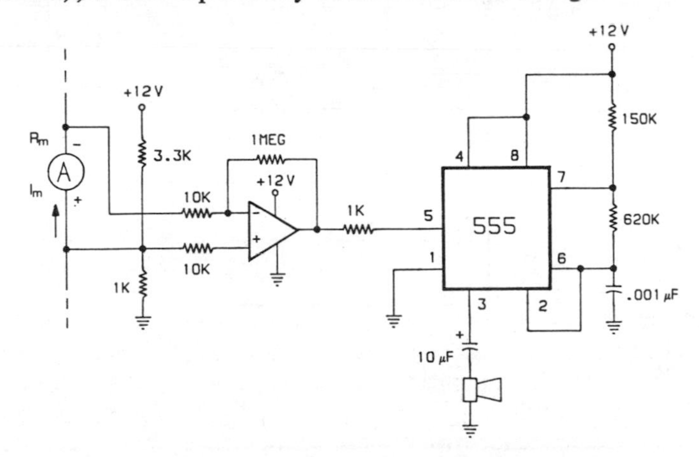

Fig. 9-13. A voltage-to-frequency converter.

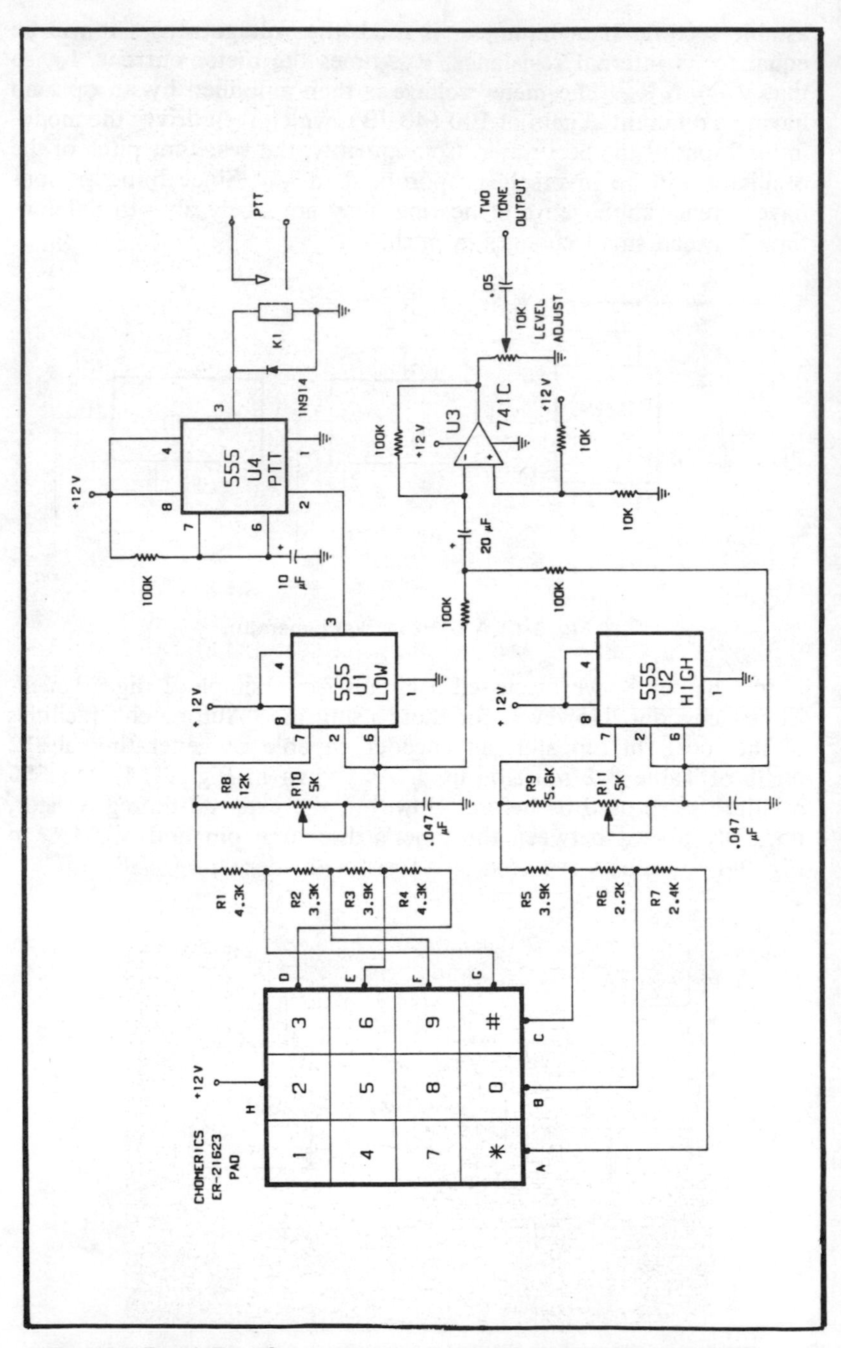

Fig. 9-14. Touch-Tone® encoder circuit capable of generating 12 digits.

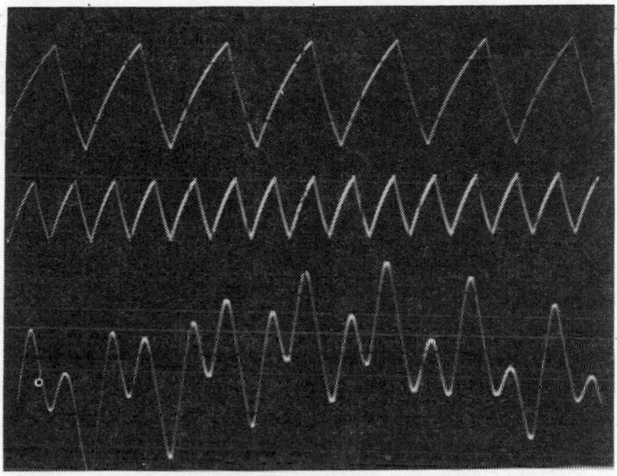

Fig. 9-15. Resulting two-tone output of op amp in Fig. 9-14.

For the low-tone oscillator (U1), using Equation 3-4 and letting $R_a = 4.3\ k\Omega = R1$, $C = 0.047\ \mu F$, and $f = 941$ Hz. Solving for R_b yields 14,164 Ω. To generate the next lower tone of 832 Hz, $R_a = R1 + R2$, so that $R2 = 3.3\ k\Omega$. For the 770-Hz tone, $R_a = R1 + R2 + R3$, making $R3 = 3.9\ k\Omega$. In a like manner, $R4 = 4.3\ k\Omega$. The high-tone oscillator (U2) is designed in a similar manner, starting with the 1477-Hz tone, by letting $R_a = R5 = 3.9\ k\Omega$, and $C = 0.047\ \mu F$. Consequently, $R6 = 2.2\ k\Omega$, and $R7 = 2.4\ k\Omega$.

For both oscillators, the output is taken from the junction of the timer's discharge and trigger pins, which produces an exponential tri-

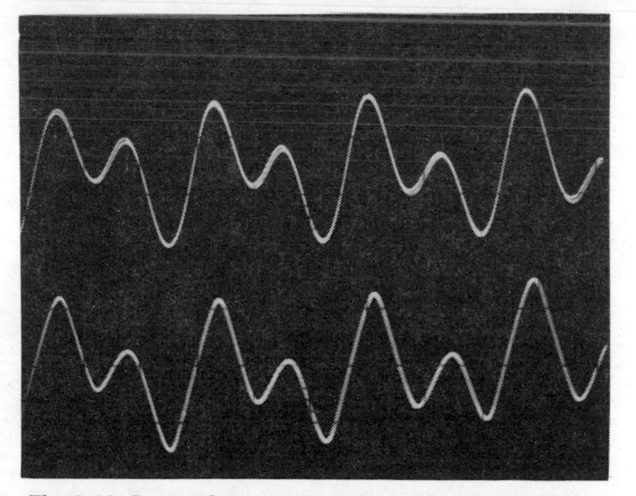

Fig. 9-16. Comparison curves representing the digit "1."

angular waveform between ⅓ and ⅔ V_{cc}. A 741 op amp then adds the output of both oscillators, coupling the resulting two-tone output (Fig. 9-15) to a 10-kΩ potentiometer that serves as an output level control.

A third timer functions as an automatic push-to-talk (PTT) control (or VOX) eliminating the need to simultaneously press the microphone switch and operate the encoder. As emphasized in Chapter 2, good thermal stability is achieved when tantalum or Mylar capacitors are used for the 0.047 μF timing capacitors.

Proper alignment is accomplished by first pressing the star (*) key and adjusting R10, so that the low-group oscillator reads 941 Hz at pin 3 of U1. Consequently, frequencies of 852, 770 and 697 Hz should be obtained to within 2% when the numbers 7, 4 and 1 are pressed respectively. For the high-tone group, press the pound (#) key and adjust R11 so that the oscillator reads 1477 Hz at pin 3 of U2. Frequencies of 1336 and 1209 Hz should then be obtained, also to within 2%, when the 0 and * keys are pressed. A comparison of the digit "1" generated by the 555 circuit and a standard Western Electric encoder is shown in Fig. 9-16.

For proper operation with a fm transmitter, the output level control (R12) should be set against a deviation meter. Otherwise, on-the-air testing will be required to set the output level. In addition, the PTT delay should be set to give about a 1-second hold time after any key is pressed.

CHAPTER 10

Experimenting With the 555 Timer

INTRODUCTION TO THE EXPERIMENTS

The following 17 experiments are designed to demonstrate the operation of the 555 timer, illustrating a few techniques and basic circuits that have already been discussed in this book. To conduct all of these experiments, you will need the following components and equipment. A BK-1 experimenter's package, sold by E&L Instruments, Inc., contains all the components to perform these experiments.

Capacitors: 0.001, 0.01, 0.1, 1, 5, 10, and 100 μF.

Fixed Resistors: 150Ω, 470Ω, 1 kΩ, 1.8 kΩ, 4.7 kΩ, 10 kΩ, 15 kΩ, 47 kΩ, 100 kΩ, 150 kΩ, 330 kΩ, 620 kΩ, 1 MΩ, and 10 MΩ. All should be either ¼ or ½ watt.

Potentiometers: 10 kΩ, and 1 MΩ.

Transistors: HEPS0013, 2N2605, or equivalent PNP type.

Diodes: 1N914

Integrated Circuits: 555 timer, 7400, 7442, 7473, and 7490.

Miscellaneous: LEDs, thermistor, photocell, normally open push-button switch, +5-V power supply, 8Ω speaker, general-purpose function generator, oscilloscope, and a vom.

The majority of the above components can be obtained from a number of suppliers that cater to experimenters. Some are listed below:

- Poly Paks
  P. O. Box 942A-2
  Lynnfield, Mass. 01940

- James Electronics
  1021-A Howard Avenue
  San Carlos, Calif. 94070
- Jade Co.
  P. O. Box 4246
  Torrance, Calif. 90510
- E&L Instruments, Inc.
  61 First Street
  Derby, Conn. 06418
- Solid State Systems, Inc.
  P. O. Box 773
  Columbia, Mo. 65201

(Note: The use of trade names and suggested suppliers of electronic components and equipment in this book does not infer an endorsement by the author but are included for the reader's convenience.)

You may find it useful to use a small selection of LR-OUTBOARDS® for the experiments. These may include:

- SK-50 Solderless Socket
- LR-1 Power Outboard
- LR-6 LED Lamp Monitor Outboard
- LR-7 Dual Pulser Outboard

Alternately, you can use the LR-25 Breadboarding Station Breadboard which, like the outboards listed above, can be obtained from E&L Instruments, Inc., Derby, Conn.

Before you set up any experiment, I recommend that you do the following:

1. Disconnect, or turn off, *all* power to the breadboard.
2. Clear the breadboard of all wires and components from previous experiments, unless instructed otherwise.
3. Carefully wire up the circuit, using whatever components that are required.
4. Check the completed circuit to make sure that it is correct. Make sure that each 555 timer and any other integrated circuit have both a ground and a +5-V connection.
5. Connect, or turn on, the power to the breadboard *last*.

GOOD LUCK.

EXPERIMENT 1

Purpose

The purpose of this experiment is to demonstrate the operation of the 555 timer as a monostable multivibrator or one-shot timer.

Schematic Diagram of Circuit

Refer to Fig. 2-1 and Fig. 10-1. This experiment can also be done using a LR-7 pulser, and a LR-6 LED lamp monitor.

Step 1

Wire the experiment according to the schematic diagram shown in Fig. 10-1, first using a 100 kΩ resistor for R_a and a 10 μF electrolytic capacitor for C. When using electrolytics, pay careful attention to the polarity! A normally open push-button switch is connected from the timer's pin 2 to ground. Or, a pulser outboard can be used with pin 2 of the timer connected to the "logic 1" output.

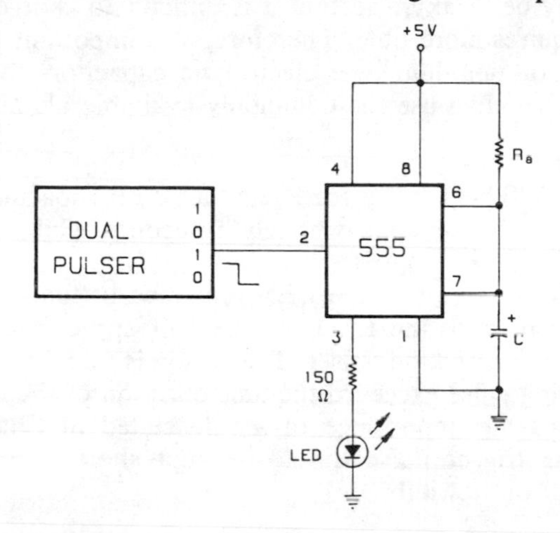

Fig. 10-1. The 555 timer connected as a monostable multivibrator.

Step 2

Using either a stopwatch or a wristwatch with a seconds hand, determine how many seconds the LED stays lit when the pulser is quickly pressed and released. Next, using the various combinations of resistors and capacitors listed below, determine the pulse duration with the pulser. Write your answers below and compare them with the equation:

$$t = 1.1(R_aC)$$

Timing Resistor	Timing Capacitor	Measured Pulse Width
100 kΩ	10 μF	
100 kΩ	100 μF	

1 MΩ	5 μF	________________
330 kΩ	10 μF	________________
330 kΩ	100 μF	________________
10 MΩ	1 μF	________________
620 kΩ	5 μF	________________
10 MΩ	5 μF	________________

The results that you obtain should be consistent to within 10% when compared with the above equation. If not, chances are that the capacitor may be "leaky," so that it is difficult to charge and consequently it requires more time. Therefore, it is important to try to use either Mylar or tantalum type electrolytic capacitors for the timing capacitor, rather than use the commonly available aluminum types.

Step 3

Now use a 330 kΩ resistor for R_a and a 10 μF capacitor for C, and press the pulser in for approximately 5 seconds while watching the LED.

You should observe that the LED remained ON until you released the pulser, or perhaps the LED remained lit approximately 3.6 seconds after you released the pulser. The reason is that the time duration of the triggering pulse exceeded the time duration of the output pulse. This illustrates the importance of a rule stated in Chapter 2—the negative-going trigger pulse should be kept short compared to the desired output pulse width.

Step 4

Using a 1 MΩ resistor for R_a and a 0.1 μF capacitor for C, determine how fast you can push and release the pulser. You should find that your limit is reached when the monostable operation becomes erratic, no matter how fast you try to press and release the pulser since this combination corresponds to a 0.11-second output pulse.

Step 5

Try a 1 MΩ resistor and a 0.001 μF capacitor in the circuit. Repeatedly press and release the pulser as quickly as you can. You should observe that it is impossible to prevent the 555 timer from acting erratic since the output pulse width will be about 1.1 msec.

EXPERIMENT 2

Purpose

The purpose of this experiment is to demonstrate the reset function of the 555 timer, connected as a monostable multivibrator.

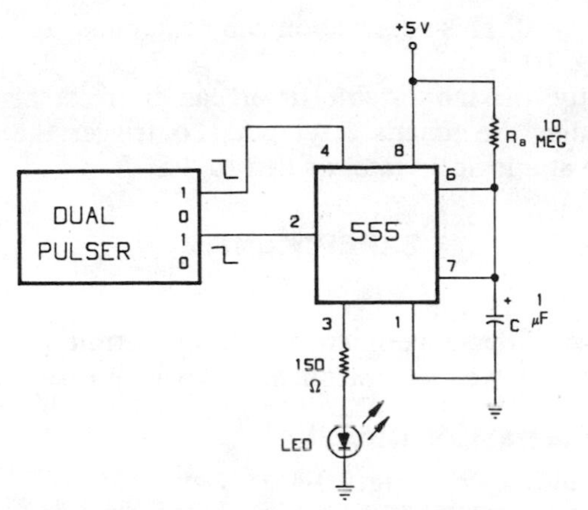

Fig. 10-2. A monostable multivibrator circuit.

Schematic Diagram of Circuit

Use the circuit shown in Fig. 10-2.

Step 1

Wire the experiment like the schematic diagram of Fig. 10-2, either using two push-button switches or two pulsers.

Step 2

Quickly press and release the pulser that is connected to pin 2 of the 555 timer. The LED should remain ON for about 11 seconds. Again, quickly press and release the pulser and, then, press and release the pulser that is connected to the timer's reset pin (pin 4). You should observe that the LED immediately goes out. It should now be obvious that the reset pin inhibits the timer when pin 4 is changed from a logic 1 to a logic 0 state, forcing the output to return

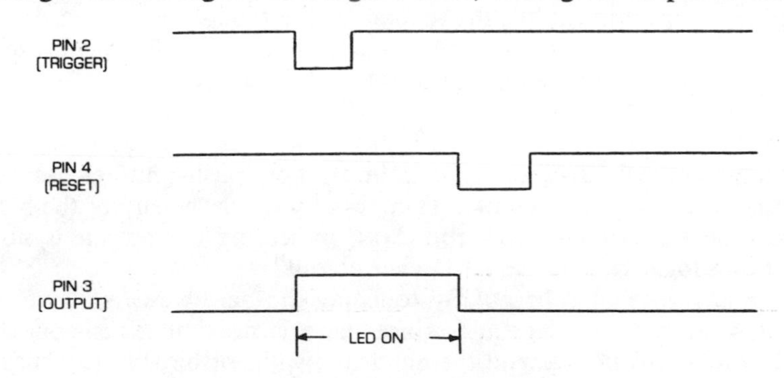

Fig. 10-3. Timing diagram.

to a LOW or logic 0. This observation can be summarized in the timing diagram of Fig. 10-3.

Consequently, the monostable timer can be reset at any time by simply grounding the timer's reset pin. To trigger the timer again, pin 4 must be at a logic 1 state, or tied to +5 V.

EXPERIMENT 3

Purpose

The purpose of this experiment is to demonstrate the operation of the 555 timer as a "touch-controlled" latch with reset.

Schematic Diagram of Circuit

Use the circuit shown in Fig. 10-4.

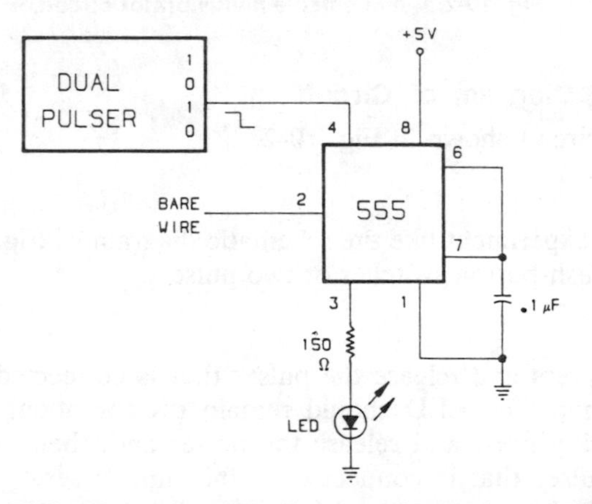

Fig. 10-4. Timer connected to allow use of a touch-control latch.

Step 1

Wire the experiment like the schematic diagram shown in Fig. 10-4. Although this circuit is the same as the one shown in Fig. 5-35, use a short piece of bare wire instead of a metal strip.

Step 2

Initially, the LED should be unlit. If not, push and release the pulser, thus resetting the timer. Then, briefly touch the end of the bare wire. The LED should now come ON, indicating the output is now HIGH or a logic 1. Does the LED ever go out?

The answer should be no! By touching the bare wire, we have, in effect, triggered the timer and, since there is no timing resistor, the time constant is infinite and the capacitor will not be able to charge. Therefore, the output will remain HIGH indefinitely.

Step 3

Now press and release the pulser that is connected to the timer's reset pin. Does the LED go out?

The answer should be yes! As demonstrated in Experiment 2, the application of a negative pulse, or grounding the reset pin, will immediately reset the timer so that the output returns to a logic 0 state or LOW.

Step 4

Now, repeat this sequence several times to demonstrate that you are able to alternatively latch the output HIGH and, then, reset the timer. As explained in Chapter 5, this circuit is very useful for touch-controlled applications. The concept of this circuit is analogous to gating and resetting a silicon-controlled rectifier.

It should now be apparent that the 555 timer, when connected as a monostable one-shot, can be very susceptible to noise or vibration.

Step 5

Reset the timer with the pulser so that the LED is unlit. Now take a pencil and gently tap the bare wire. Does the LED light up?

I have found that the LED will light up most of the time, indicating that the circuit is susceptible to vibration. However, the standard monostable circuit, as discussed in Experiment 1, is relatively immune to noise and vibration.

EXPERIMENT 4

Purpose

The purpose of this experiment is to demonstrate the use of the control pin to vary the duration of the timer's output pulse.

Schematic Diagram of Circuit

Fig. 10-5 shows the circuit to be used in this experiment.

Step 1

Wire the experiment as shown in the schematic diagram of Fig. 10-5, preferably using a 1 MΩ potentiometer for R1 instead of fixed resistors. If a potentiometer is not available, a number of fixed resistors can be used instead.

Step 2

First, temporarily disconnect the 10 kΩ resistor that is connected between potentiometer R1 and the timer's control pin (pin 5). Using either a stopwatch or a wristwatch with a seconds hand, determine

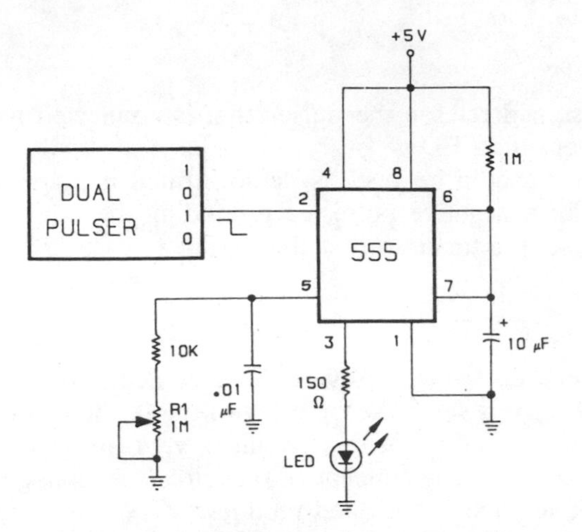

Fig. 10-5. Using the control pin of the timer to vary the output pulse.

how long the LED stays lighted when the pulser is quickly pressed and released. It should be approximately 11 seconds, depending on the tolerance of the 1 MΩ resistor and the quality of the 10 μF capacitor.

Step 3

Momentarily disconnect the power supply and reconnect the 10 kΩ resistor. Then, adjust the potentiometer to about 100 kΩ (using a vom), or use a 100 kΩ fixed resistor, and apply the power. Using the pulser, note how long the LED stays lighted. When I performed this, the LED remained on for about 10 seconds.

Step 4

Disconnect the power and adjust the potentiometer to about 10 kΩ, or use a 10 kΩ fixed resistor instead. Reconnect the power and use the pulser, noting the time that the LED stays on. You should obtain a time of about 8.5 seconds.

Step 5

Consequently, as you decrease the potentiometer's resistance further, the LED will remain on for a shorter time. Using the graph of Fig. 2-9, try varying the potentiometer's resistance to give different output pulse durations. You should obtain pulse durations from about 7 seconds to 11 seconds.

EXPERIMENT 5

Purpose

The purpose of this experiment is to demonstrate the operation of the 555 timer as an astable multivibrator or clock.

Schematic Diagram of Circuit

The circuit for this experiment is shown in Fig. 10-6.

Step 1

Wire the experiment as shown in the schematic diagram using a 1 kΩ resistor for R_a, a 1 MΩ resistor for R_b, and a 1 μF capacitor for C. Be sure to connect the timer's reset pin (pin 4) to +5 volts.

Step 2

Connect the power to the breadboard and observe that the LED periodically flashes on and off. With the aid of a stopwatch or wristwatch, count the number of times the LED is lighted over a 1-minute period.

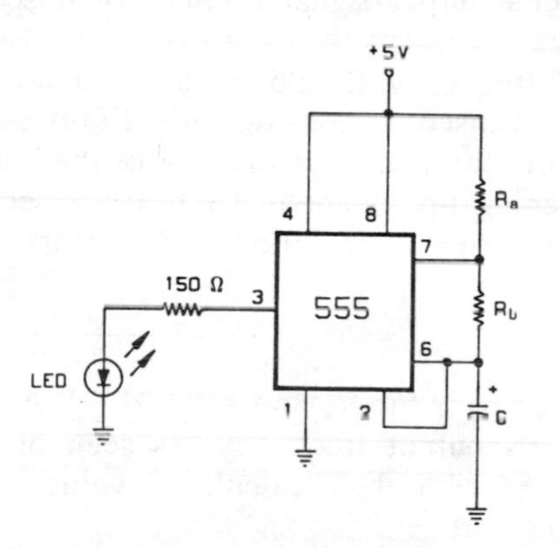

Fig. 10-6. The 555 timer as an astable multivibrator.

You should count about 44 pulses per minute, which is the same as a frequency of 0.73 Hz and is slow enough for the eye to follow. This can then be compared with the value calculated from Equation 3-4, or

$$f = \frac{1.443}{(R_a + 2R_b)C}$$

which, with the resistance and capacitance values used in Step 1, gives a calculated value of 0.716 Hz.

Step 3

Disconnect the power to the breadboard and connect a 5 μF capacitor in place of the 1 μF capacitor. First calculate the expected frequency or the number of pulses you should observe in 1 minute. You should have calculated a value of 0.144 Hz or 8.65 pulses per minute.

Step 4

Reconnect the power to the breadboard, and count the number of pulses in a 1-minute period. Does your answer reasonably agree with the value calculated in Step 3?

Step 5

The rest of this experiment can be accomplished only if you have an oscilloscope. Connect a 0.001 μF capacitor for the timing capacitor and connect the oscilloscope's probe to pin 3 of the 555 timer.

Step 6

Reconnect the power to the breadboard and observe the LED. You should observe that the LED appears to be continuously ON. Now observe the timer's output signal on the oscilloscope. What is the output frequency?

The observed frequency should be approximately 720 Hz, or a full cycle every 1.4 msec. Therefore, the LED is actually flashing on and off at a rate of 720 times a second. Since the human eye can only see things happening up to about 16 times a second, the LED is flashing faster than the eye can follow and appears to be continuously ON.

Step 7

Using a 1 kΩ resistor for R_a and a 0.001 μF capacitor for C, determine the timer's output frequency, as seen on the oscilloscope, and compare them with their calculated values for the resistance values listed below:

Timing Resistor R_b	Observed Frequency	Calculated Frequency
330 kΩ	________	________
100 kΩ	________	________
15 kΩ	________	________
10 kΩ	________	________
4.7 kΩ	________	________
1.8 kΩ	________	________

You may have observed erratic behavior on the oscilloscope when using a 1.8 kΩ resistor for R_b, which corresponds to an astable frequency of about 314 kHz. Since the maximum astable frequency possible with a 555 timer is about 300 kHz, the output will become erratic or nonexistent if an output frequency greater than 300 kHz is attempted.

EXPERIMENT 6

Purpose

The purpose of this experiment is to demonstrate the effect of the timing resistors on the duty cycle while using the 555 timer as an astable multivibrator. In order to fully appreciate this experiment, an oscilloscope is required.

Schematic Diagram of Circuit

The circuit shown in Fig. 10-7 will be used in this experiment.

Step 1

For this experiment, an oscilloscope is required. Wire the experiment as shown in the schematic diagram of Fig. 10-7, first using a 1 kΩ resistor for R_a, a 100 kΩ resistor for R_b, and a 0.01 μF capacitor for C.

Step 2

Apply power to the breadboard and apply the oscilloscope's probe to pin 3 of the timer. What is the output frequency and duty cycle?

You should have determined that the square wave period is about 1.4 msec, giving a frequency of 720 Hz. The duty cycle is the ratio of the time period that the output is HIGH, or at +5 V, to the time for one complete cycle. In this case, the duty cycle is very nearly 50%. Compare this result with Equation 3-5.

Step 3

Now replace the 1 kΩ resistor (R_a) with a 100 kΩ resistor, and the 100 kΩ resistor (R_b) with a 47 kΩ resistor. What is the frequency and duty as now seen on the oscilloscope?

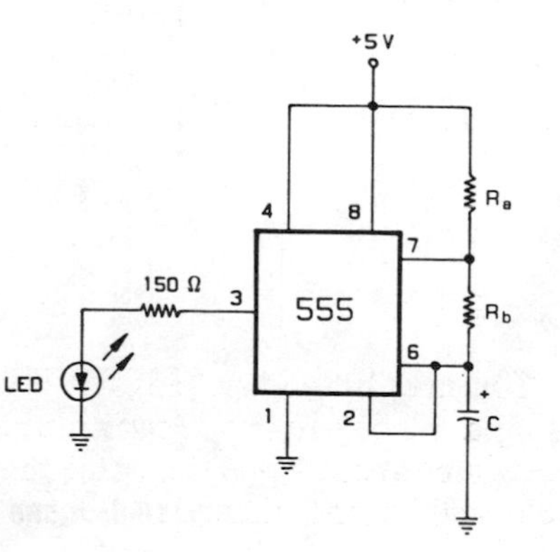

Fig. 10-7. The effect of timing resistors on the duty cycle of an astable multivibrator.

The frequency should still be approximately 720 Hz, but the duty cycle has changed from 50% to about 75%. The output signal now has a HIGH period that is three times greater than the OFF period. Although the frequency has essentially remained the same, the duty cycle has been increased by 50%.

Step 4

With the aid of Fig. 3-3, try different resistor combinations to try to produce a wide range of duty cycles. In this experiment, you should have concluded that it is difficult to vary the duty cycle while keeping the output frequency constant. A more practical circuit is described in the next experiment.

EXPERIMENT 7

Purpose

The purpose of this experiment is to demonstrate that it is possible to vary the timer's duty cycle while holding the output frequency relatively constant.

Schematic Diagram of Circuit

Use the circuit shown in Fig. 10-8 for this experiment.

Step 1

For this experiment, an oscilloscope is required. Wire the experiment as shown in the schematic diagram of Fig. 10-8 and connect the oscilloscope probe to pin 3 of the timer.

Step 2

Apply power to the breadboard. What is the output frequency and the duty cycle? Your answer should be approximately 106 Hz since

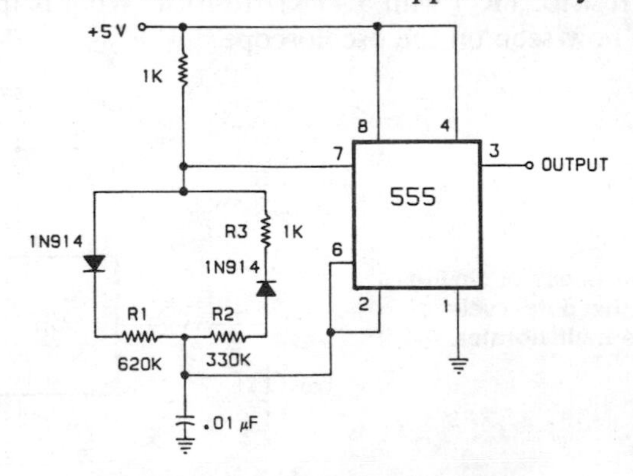

Fig. 10-8. A 555 timer circuit.

the time duration for one complete cycle is 9.4 msec. The time duration that the output is HIGH or at +5 volts is about 3.3 msec, so that the duty cycle is 3.3 msec/9.4 msec, or 35%.

For this circuit, the equation for determining the output frequency from the circuit components is different than the one given in Experiment 5. The output frequency is given by the expression

$$f = \frac{1}{(R1 + R2 + R3)C}$$

and the expected frequency is 105 Hz.

Step 3

Disconnect the power and interchange the 330 kΩ and 620 kΩ resistors. Then reconnect the power to the breadboard and observe the output waveform on the oscilloscope. What, if anything, happened?

You should have observed that the duty cycle increased from 35% to 65%, but there was *no* change in the output frequency! From noticing the difference between Step 2 and Step 3, you should have concluded that the duty cycle depends on the position of the *larger* resistor (620 kΩ). To calculate the duty cycle for this circuit, use

$$D = \frac{R2}{R1 + R2}$$

Therefore, if R2 is larger than R1, the duty cycle will be greater than 50%. Also, is R2 is less than R1, the duty cycle will be less than 50%. If both resistors are equal, the duty cycle will be exactly 50%.

Step 4

Disconnect the power from the breadboard and replace the 330 kΩ and 620 kΩ resistors with a 1 MΩ potentiometer as shown in Fig. 10-9.

Step 5

Reconnect the power to the breadboard. Now vary the potentiometer's resistance from one extreme to the other. What do you observe about the duty cycle and output frequency as the potentiometer resistance is varied?

You should have noticed that the duty varied from almost 3% to about 97% without any change in frequency, depending upon the

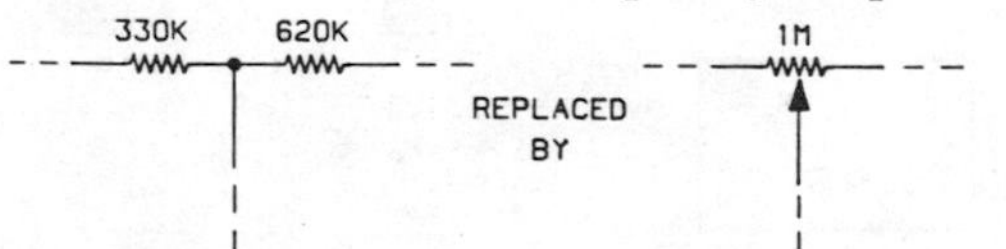

Fig. 10-9. Substituting a potentiometer for two fixed resistors.

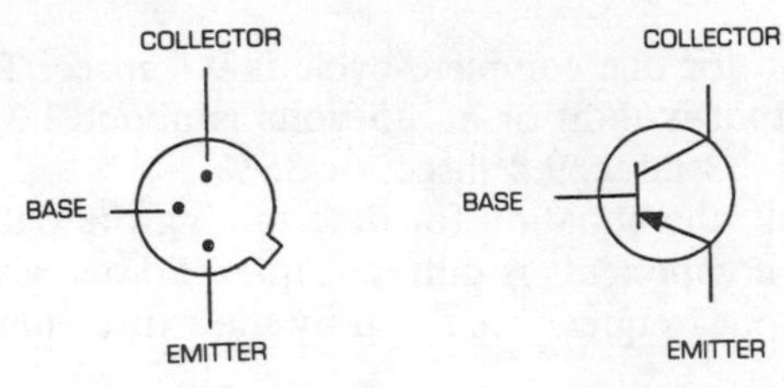

Fig. 10-10. Transistor pin configuration.

potentiometer's setting. This circuit technique is very useful for smoothly changing the timer's duty cycle, with only a single adjustment and without affecting the output frequency.

EXPERIMENT 8

Purpose

The purpose of this experiment is to demonstrate the operation of the 555 timer as a missing pulse detector.

Transistor Pin Configuration

If you haven't worked with transistors before, you may permanently damage the device if you accidentally connect the leads incorrectly. As seen from the bottom, the transistor's leads are arranged in the order shown in Fig. 10-10.

Schematic Diagram of Circuit

Use the missing pulse detector circuit shown in Fig. 10-11.

Step 1

Wire the experiment as shown in the schematic diagram of Fig. 10-11. Be careful that you connect the transistor properly! Also, don't

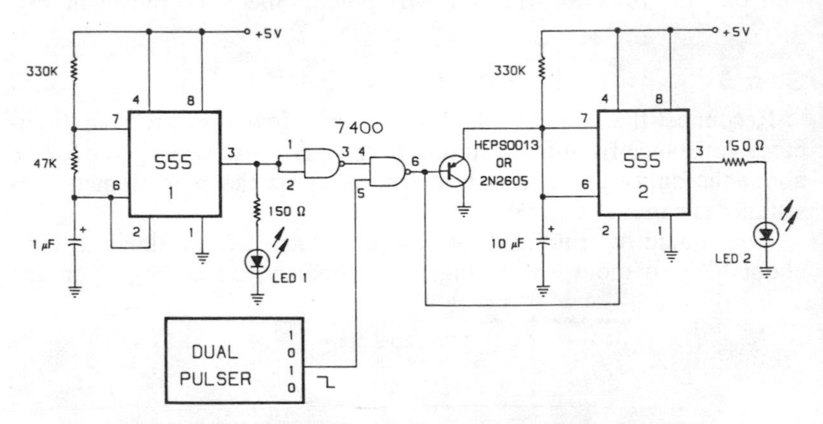

Fig. 10-11. A missing pulse detector circuit using a 555 timer.

forget the power connections to the 7400 TTL NAND gate, since they are generally omitted on schematic diagrams. Pin 14 is connected to +5 V, and pin 7 is grounded.

Step 2

Connect the power to the breadboard. LED 1 should alternately flash ON and OFF at a very slow rate. Is LED 2 ON or OFF?

LED 2 should be continuously ON, since the output pulses from timer #1 are constantly triggering timer #2. Since the time between clock pulses is less than the monostable ON time of about 3 seconds, the output of the second timer will remain HIGH and LED 2 will be lighted.

Step 3

Now hold the pulser in. Does LED 2 go out?

The LED should extinguish approximately 3 seconds after the pulser is pushed in. A logic 0 from the pulser to one of the inputs of the second 7400 NAND gate blocks or inhibits the clock pulses from triggering the second timer. Since no pulses are permitted to trigger timer #2, the timer consequently completes its monostable cycle, at which time the output goes LOW. This sequence is best understood by following the timing diagram shown in Fig. 10-12.

Step 4

Now release the pulser. What happens to LED 2?

LED 2 should immediately come ON. The output pulses from the first timer are again triggering the second timer. The second timer then serves to detect either the presence or absence of a recurring pulse, and the LED acts as the visual indicator.

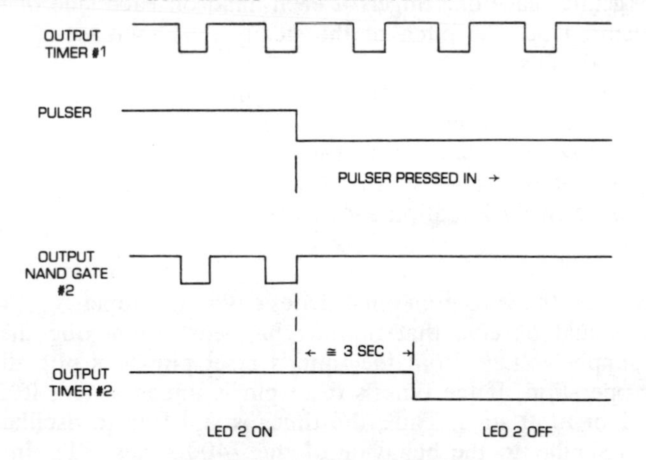

Fig. 10-12. Timing diagram.

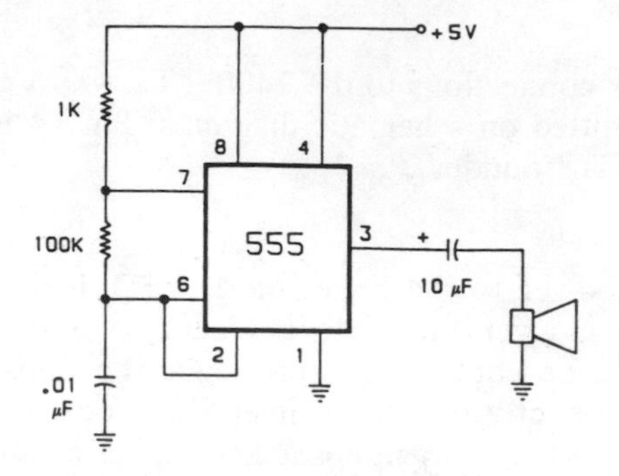

Fig. 10-13. An audio oscillator.

EXPERIMENT 9

Purpose

The purpose of this experiment is to connect the 555 timer to function as an audio oscillator.

Schematic Diagram of Circuit

Fig. 10-13 shows the circuit for an audio oscillator.

Step 1

Wire the experiment as shown in the schematic diagram of Fig. 10-13. Then connect the power to the breadboard. You should then hear a steady tone (about F#) from the speaker. If not, carefully check your wiring.

Step 2

Now gently place one finger of each hand on each side of the 100 kΩ resistor. Does the pitch of the steady tone increase or decrease when you do this?

You should find that the tone's pitch increases. This is because the body's resistance, which can be as high as 200 kΩ between the hands, is now in parallel with the 100 kΩ resistor. Consequently, this equivalent resistance is now lower causing the frequency to increase, as given by Equation 3-4.

Step 3

Disconnect the wire from pin 4. Does anything happen?

You should observe that nothing happens! Removing the +5-V power-supply voltage from the timer's reset pin does not affect the timer's operation. If the timer's reset pin is unconnected, it assumes a logic 1 or HIGH state. Thus, the timer is still free to oscillate. This action is similar to the behavior of the 7400-series TTL integrated circuits; that is, the logic state for any unconnected input pin is always a logic 1.

Step 4

Now ground pin 4. What happens?

When pin 4 is grounded, there should be no tone, and the timer is disabled. Removing the timer's reset pin from ground then activates the timer again. This "gating" technique is very useful for alarm circuits.

EXPERIMENT 10

Purpose

The purpose of this experiment is to wire two 555 timers to create a 2-note tone.

Schematic Diagram of Circuit

Use the circuit shown in Fig. 10-14 for this experiment.

Step 1

Wire the experiment like the schematic diagram shown in Fig. 10-14, using a pair of 555 timers. If desired, a single 556 timer can be used instead, but you will have to refer to Fig. 1-2 for the proper pin connections.

Step 2

Apply power to the breadboard. What do you hear?

You should now hear a distinctive 2-note sound. The frequency of the second timer, operating around 400 Hz, is changed at a rate of approximately 10 times a second by the first timer.

Step 3

With the power still connected to the breadboard, remove the 100 kΩ resistor from the circuit. You should hear only a single 400-Hz

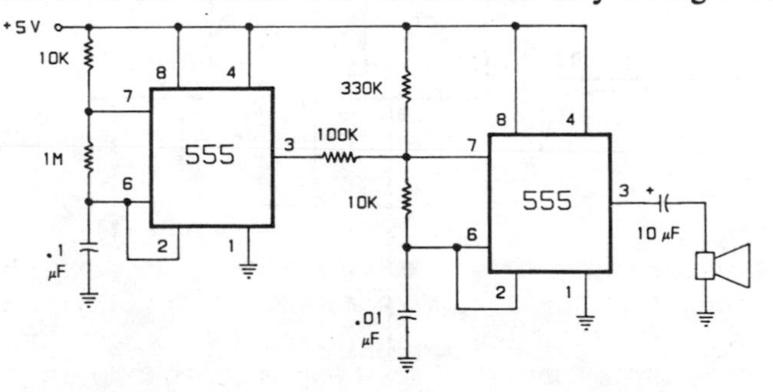

Fig. 10-14. Two 555 timers wired to create a 2-note tone.

tone. Each time the output of the first timer goes HIGH, the 100 kΩ resistor is effectively placed in parallel with the 330 kΩ resistor, giving an alternate frequency of about 1.5 kHz.

Step 4

Disconnect the power from the breadboard. Replace the 0.1 μF timing capacitor (on the first timer) with a 1 μF capacitor to give a slower rate. Then reconnect the 100 kΩ resistor as before. Apply the power and you should now be able to more easily distinguish the 400-Hz and 1.5-kHz tones.

EXPERIMENT 11

Purpose

The purpose of this experiment is to cascade, or connect in series, a pair of 555 timers to sequentially light a pair of LEDs.

Schematic Diagram of Circuit

First refer to the schematic diagram given in Experiment 1. If you have not already performed Experiment 1, do so now before attempting this experiment.

Step 1

Wire the experiment as shown in the schematic diagram of Fig. 10-15. There are two 555 timers, each wired to function as monostable vibrators. The output of the first timer is connected to the second timer by a 0.01 μF capacitor and 10 kΩ resistor.

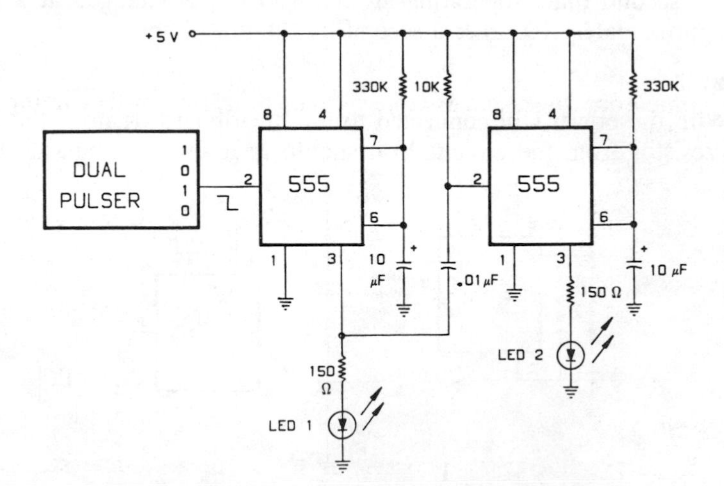

Fig. 10-15. Cascaded 555 timers.

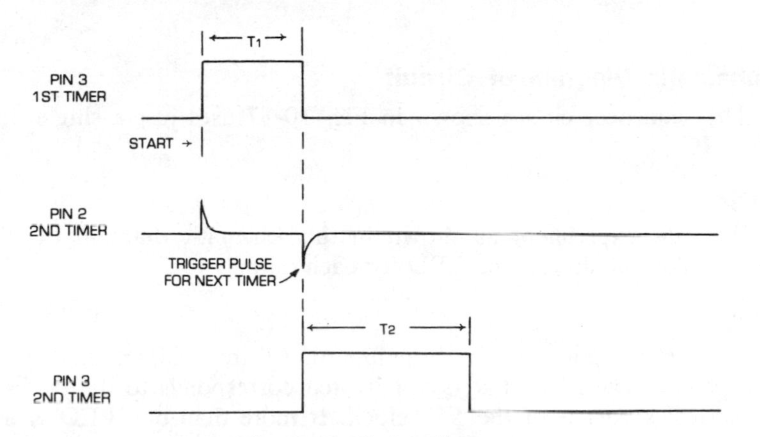

Fig. 10-16. Timing diagram.

Step 2

Connect the power to the breadboard. Both LEDs should be unlighted. If not, perhaps one of the timers was triggered when the power was applied to the circuit. Wait several seconds for both LEDs to be OFF.

Step 3

Now, with both LEDs in the OFF state, quickly press and release the pulser. What happens to both LEDs?

LED 1 should immediately light up. After approximately 3 seconds, LED 1 goes out and LED 2 is lit for 3 seconds. At the completion of the first timer's monostable period, the output of the first timer goes LOW. This HIGH-to-LOW transition is then differentiated by the 0.01 μF capacitor and 10 kΩ resistor to produce the necessary short negative-going pulse to trigger the next timer. When the output of the first timer triggers the next timer, its output will remain HIGH until the completion of its monostable period, as shown in the timing diagram of Fig. 10-16.

This sequence of events is the same as if you had two independent timers with two pulsers. But, in this experiment, we let the output of one timer automatically trigger the next one. Consequently, we can cascade any number of timers to produce a sequence of output commands. In addition, the time duration between sequences can be individually adjusted by varying the timing components of the individual timer.

EXPERIMENT 12

Purpose

The purpose of this experiment is to create a 10-line sequencer, using a single 555 timer.

Schematic Diagram of Circuit

The sequencer circuit shown in Fig. 10-17 uses just a single 555 timer IC.

Step 1

Wire the experiment as shown in the schematic diagram of Fig. 10-17. You should use one LED for each output.

Step 2

Connect the power to the breadboard. All the LEDs should be lit except one. The LED that is not lighted corresponds to the decimal numerical sequence of the 555 clock. If more than one LED is unlighted at the same time, check your wiring for possible errors.

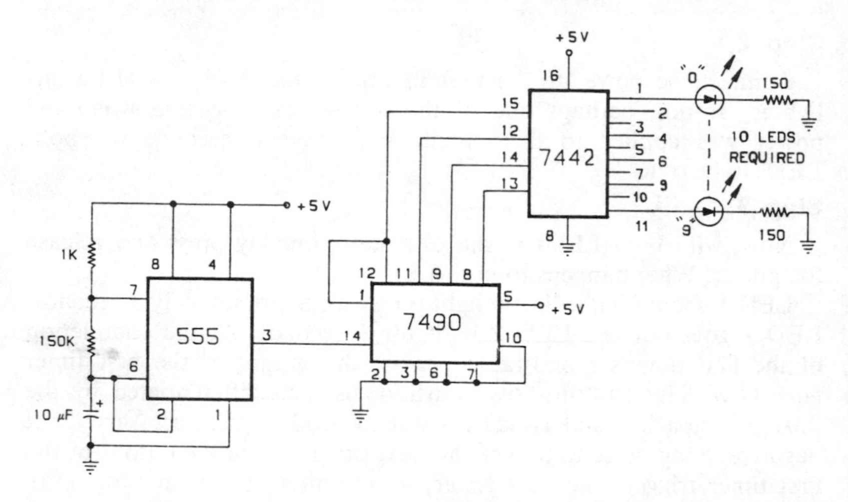

Fig. 10-17. A 10-line sequencer circuit.

The purpose of the 7442 4-line-to-10-line decoder is to convert the 4-bit binary word from the 7490 integrated circuit into one of ten outputs. If you are not familiar with the term "binary word," or the operation of decoders, refer to Book 1 of *Logic & Memory Experiments Using TTL Integrated Circuits* in the Blacksburg Continuing Education Series. For any one of the ten decimal numbers between 0 and 9, only *one* of the outputs of the 7442 chip is at a logic 0 or LOW state; the remaining nine outputs are all at logic 1. This is summarized in Table 10-1. Consequently, the output is continuously sequenced through the 10 outputs, one at a time. However, unlike the previous experiment, the outputs will change states at the same rate.

Table 10-1. Truth Table for Experiment 12

Clock Pulse	Input D	C	B	A	Outputs 0	1	2	3	4	5	6	7	8	9
0	0	0	0	0	0	1	1	1	1	1	1	1	1	1
1	0	0	0	1	1	0	1	1	1	1	1	1	1	1
2	0	0	1	0	1	1	0	1	1	1	1	1	1	1
3	0	0	1	1	1	1	1	0	1	1	1	1	1	1
4	0	1	0	0	1	1	1	1	0	1	1	1	1	1
5	0	1	0	1	1	1	1	1	1	0	1	1	1	1
6	0	1	1	0	1	1	1	1	1	1	0	1	1	1
7	0	1	1	1	1	1	1	1	1	1	1	0	1	1
8	1	0	0	0	1	1	1	1	1	1	1	1	0	1
9	1	0	0	1	1	1	1	1	1	1	1	1	1	0
10	0	0	0	0	0	1	1	1	1	1	1	1	1	1
11	0	0	0	1	1	0	1	1	1	1	1	1	1	1
Etc.	Etc.				Etc.									

EXPERIMENT 13

Purpose

The purpose of this experiment is to demonstrate the operation of the 555 timer as a Schmitt trigger.

Schematic Diagram of Circuit

The circuit for this experiment is shown in Fig. 10-18.

Step 1

Wire the experiment as shown in the schematic diagram of Fig. 10-18. For this experiment, you will need a function generator and an oscilloscope that has a dual trace. Connect one probe to the circuit's input, and the other probe to pin 3 of the timer.

Step 2

Connect the power to the breadboard. Adjust the output of the function generator to give a 1.5-V peak (3-V peak-to-peak) 1-kHz

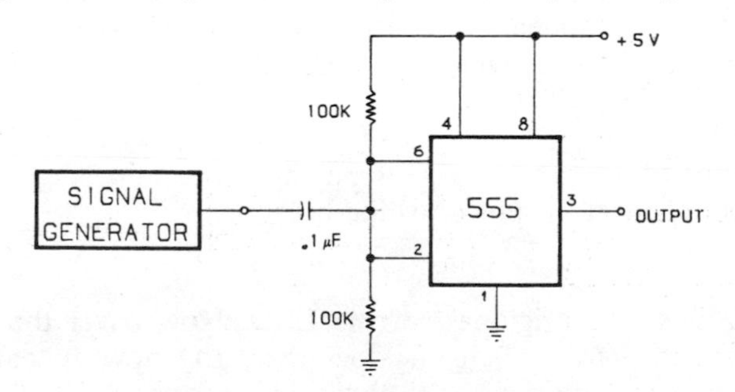

Fig. 10-18. A Schmitt trigger.

sine wave as your input signal. At what voltage on the input sine wave does the output change from HIGH to LOW? From LOW to HIGH?

The timer's output changes from HIGH to LOW when the input signal equals +0.8 V, going positive from its negative-peak voltage of −1.5 V. Also, the timer's output changes from LOW to HIGH when the input signal equals −0.8 V, going negative from its positive-peak voltage of +1.5 V, as illustrated in the diagram of Fig. 10-19. These two voltage levels, when added to the +2.5-V bias set by the two 100 kΩ resistors, are identically equal to the internal comparator's set points of ⅔ V_{cc} (3.3 V) and ⅓ V_{cc} (1.7 V). What a coincidence.

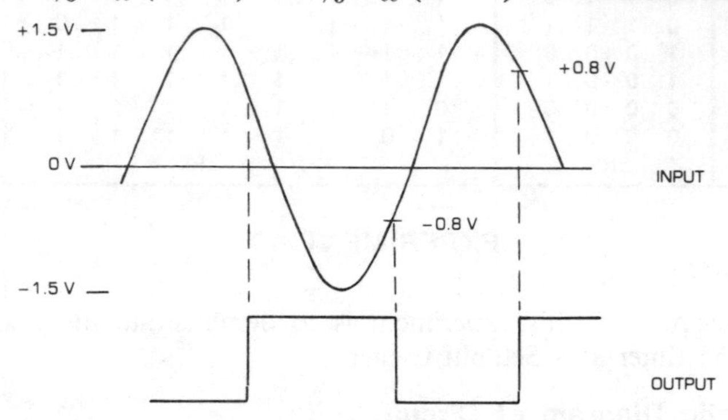

Fig. 10-19. Voltage points on the input sine wave.

EXPERIMENT 14

Purpose

The purpose of this experiment is to demonstrate the use of a photocell with a 555 timer to control the brightness of a LED display.

Schematic Diagram of Circuit

As shown in the schematic of Fig. 10-20, this circuit uses a photocell. I used a Type CL703L made by Clairex, which can be obtained from Poly Paks for about 50¢ each. The BK-1 experimenter's package sold by E&L Instruments, Inc. uses this type. However, almost any other type can be used with similar results.

Step 1

Wire the experiment as shown in the schematic diagram of Fig. 10-20 and connect the power to the breadboard.

Step 2

First, notice the brightness of the LED. Now cover the photocell so it is shielded from all light. Does the LED now appear brighter or dimmer?

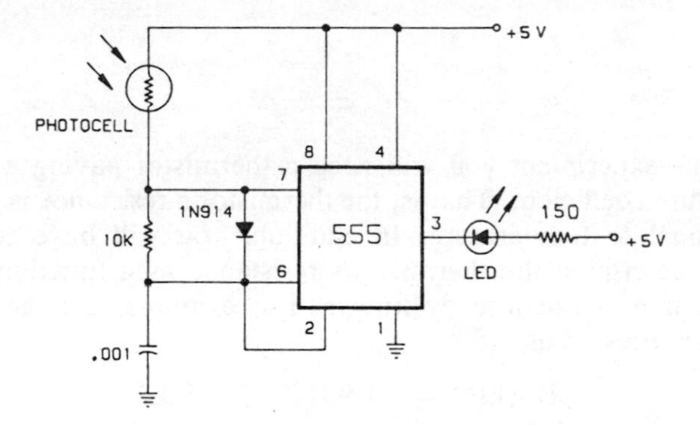

Fig. 10-20. A brightness control using a photocell and a 555 timer.

If you wired the circuit correctly, the LED should be brighter! The timer's output duty cycle has now decreased, causing the LED to remain ON longer during each cycle.

Step 3

Now uncover the photocell. The LED should be less bright since the timer's duty cycle is greater, and the LED is ON for a shorter duration than before.

My photocell had a resistance of about 1 kΩ when fully exposed to light. When covered, the resistance increased to about 30 kΩ. The duty cycle, therefore, varied from approximately 16% to about 75% when going from totally dark to fully lighted conditions.

EXPERIMENT 15

Purpose

The purpose of this experiment is to demonstrate the operation of the 555 timer with a thermistor.

Schematic Diagram of Circuit

The circuit for this experiment is shown in the schematic given in Fig. 10-21.

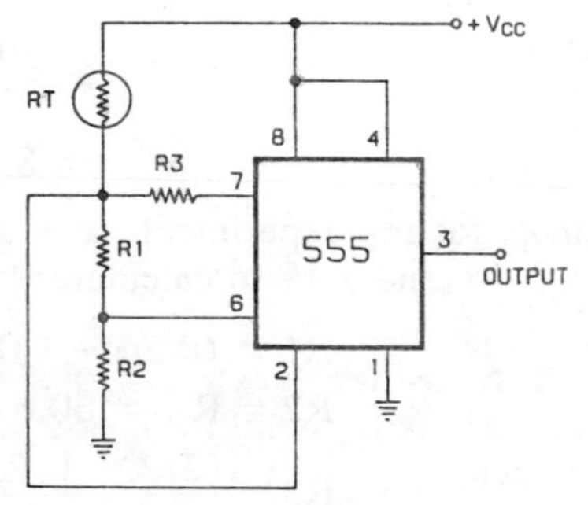

Fig. 10-21. A circuit containing a 555 timer and a thermistor.

Step 1

For this experiment you will need a thermistor having a negative temperature coefficient. That is, the thermistor's resistance is inversely proportional to temperature. In addition, you will have to experimentally determine the thermistor's resistance as a function of temperature, if it is not already known. For example, the thermistor I used was expressed as:

$$R_T(k\Omega) = -1.931(T_C) + 90.2$$

where,

T_C is the temperature expressed in °C.

For temperatures given in °F, the equation is:

$$R_T(k\Omega) = -1.064(T_F) + 124.4$$

Step 2

In this experiment, we want to be able to detect when the thermistor's temperature falls outside the 60-110 °F range. Once you know your thermistor's characteristics, first determine its resistance at 60 °F (R_{TC}). In my case, this value was:

$$R_{TC} = (-1.064)(60) + 124.4$$
$$= 60.6\ k\Omega$$

Then determine its resistance at 110 °F. Again, in my case,

$$R_{TH} = (-1.064)(110) + 124.4$$
$$= 7.4\ k\Omega$$

As explained in Chapter 5 (section on Temperature Measurement and Control), we then determine the ratio

$$\alpha = \frac{R_{TC}}{R_{TH}}$$

or

$$= \frac{60.6\ k\Omega}{7.4\ k\Omega}$$
$$= 8.19$$

Since, for my experiment, α is greater than 2, we use Equations 5-17, 5-18, and 5-19 to calculate R1, R2, and R3, or

$$R1 = (0.5\alpha - 1)R_{TH} = 22.9\ k\Omega$$
$$R2 = R_{TC} = 60.6\ k\Omega$$
$$R3 = \frac{(3\alpha^2 - 1)R_{TH}}{4\alpha - 2} = 48.2\ k\Omega$$

I used standard resistor values of 22 kΩ, 62 kΩ, and 47 kΩ without much loss of accuracy. If the ratio α is less than 2, Equations 5-20 through 5-22 are to be used.

Step 3

After you have determined the required values for R1, R2, and R3 based upon your thermistor's characteristics, wire the circuit shown in the schematic diagram of Fig. 10-21, and connect the power to the breadboard. Momentarily ground pin 2. The LED should now be ON. If not, either check your wiring or your calculations. Then, place a lighted match close to the thermistor for a couple of seconds. What happens to the LED?

The LED should go out after a short time since the temperature is higher than 110 °F.

Step 4

Now place a piece of ice on the thermistor. What happens to the LED?

The LED should remain OFF for a short time and then come ON.

EXPERIMENT 16

Purpose

The purpose of this experiment is to demonstrate the operation of the 555 timer as a simple analog frequency meter or tachometer.

Schematic Diagram of Circuit

Refer to the schematic diagram in Fig. 10-22.

Step 1

Wire the experiment as shown in the schematic diagram, using either a 1-mA meter or a vom. In either case, be extremely careful of the meter's polarity; otherwise, you may permanently damage the meter!

Step 2

First use a 100 kΩ resistor for R_a. The first 555 timer will be our source of input pulses for our tachometer. Connect the power to the breadboard and adjust the 10 kΩ potentiometer so that the meter reads full scale, or 1 mA. Since the input pulses are approximately 1 kHz, a full-scale meter reading will then be equal to 1000 Hz.

Step 3

For each value of R_a indicated below, which changes the frequency of the source pulses, complete the table with your experimental results.

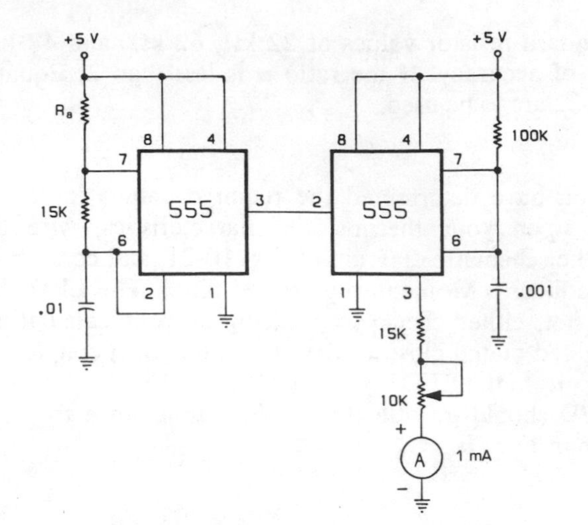

Fig. 10-22. A simple tachometer.

R_a	Meter Reading, mA
150 kΩ	________
330 kΩ	________
620 kΩ	________
1 MΩ	________

You should have obtained approximate readings of 0.8, 0.4, 0.2, and 0.15 mA respectively, which corresponds roughly to input frequencies of 800, 400, 200, and 150 Hz.

Step 4

If you are fortunate enough to have access to a digital multimeter that measures current, repeat Step 3 using the digital meter. When I performed this experiment, I compared the values obtained with a digital meter simultaneously against a $2,000 frequency counter, and the results were as listed below:

R_a	f (calculated)	f (measured)	Digital Meter
100 kΩ	1,100 Hz	1,017 Hz	1.017 mA (adjusted)
150 kΩ	802 Hz	775 Hz	0.773 mA
330 kΩ	401 Hz	381 Hz	0.376 mA
620 kΩ	222 Hz	202 Hz	0.196 mA
1 MΩ	140 Hz	120 Hz	0.110 mA

As you can see, our simple tachometer did remarkably well considering that its performance was compared against a $2,000 frequency counter. The differences between the calculated and measured values for the source frequencies are attributed to the tolerances of the components used.

EXPERIMENT 17

Purpose

The purpose of this final experiment is to demonstrate the use of the 555 timer to make a simple "heads or tails" game.

Schematic Diagram of Circuit

The schematic diagram for this experiment is shown in Fig. 10-23.

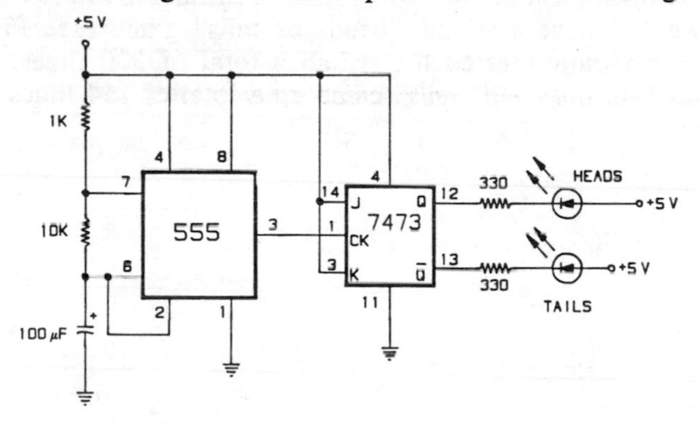

Fig. 10-23. A simple "heads or tails" game circuit.

Step 1

Wire the experiment as shown in the schematic diagram, and label one of the LEDs "heads" and the other "tails." Alternatively, you can use red and green colored LEDs.

Step 2

Connect the power to the breadboard. You should now observe that both LEDs alternately flash ON and OFF at a very slow rate. If not, check your wiring, especially the connections to the 7473 integrated circuit.

Step 3

Now, disconnect the power from the breadboard and replace the 100-μF capacitor with a 1-μF capacitor. Can you now tell which LED is ON at any given time?

Your answer should be, "No." Since each LED is now flashing ON and OFF approximately 35 times each second, both LEDs appear to be ON simultaneously and it is therefore impossible to determine which LED is ON at any given time. Now, disconnect the wire from pin 3 of the 555 timer. Now, only one LED is lit—the LED that happened to be ON when the wire was disconnected.

Step 4

Now briefly touch the wire again to pin 3 of the timer. Does the LED that was ON before again light up?

The same LED may, or may not, again be lit, depending on the exact moment the wire was disconnected. Thus, the probability that any one of the two LEDs will be lit is a 50-50 choice. By using a normally open push-button switch instead of having to disconnect the wire, we will have a simple "heads or tails" game (see Fig. 6-1). When I randomly pressed the switch a total of 300 times, "heads" came up 146 times and "tails" came up a total of 154 times.

Appendix

References

The following is a list of the articles which were used in the preparation of this book, and which have been grouped according to the organization of this book.

GENERAL INFORMATION

Hnatek, E. R. "Put the IC Timer to Work in a Myriad of Ways," *EDN,* March 5, 1973, p. 54.

Jung, W. G. "The IC Time Machine," *Popular Electronics,* November, 1973, p. 54.

Jung, W. G. "Applications for the IC Time Machine," *Popular Electronics,* January, 1974, p. 72.

Robbins, M. S. "Making Noises With the 555 IC Timer," *Popular Electronics,* July, 1974, p. 50.

Sandberg, B. "State Diagrams for a 555 Timer Aid Development of New Applications," *Electronic Design,* August 16, 1976, p. 100.

Scott, R. F. "555 Timer IC Applications," *Radio Electronics,* Part I–February, 1976, p. 40; Part II–March, 1976, p. 62; Part III–September, 1976, p. 63.

Schopp, W. S. "Versatile IC Timer," *Popular Electronics,* July, 1973, p. 98.

MONOSTABLE CIRCUITS

Klinger, A. R. "Integrated Timer Operates as Variable Schmitt Trigger," *Electronics,* October 25, 1973, p. 208.

Lalitha, M. K. and P. R. Chetty. "Variable-Threshold Schmitt Trigger Uses 555 Timer," *EDN,* September 20, 1976, p. 112.

Lickel, K. "Compensating the 555 Timer for Capacitance Variations," *Electronics,* February 6, 1975, p. 96.

Paiva, M. O. "Start a Logic in the Proper Mode When the Power is Turned on or Interrupted," *Electronic Design,* August 16, 1976, p. 98.

Satyanarayana, P. "Switched Current Source Increases IC Timer Delay," *Electronic Engineering,* October, 1975, p. 21.

ASTABLE CIRCUITS

Althouse, J. "IC Timer, Stabilized by Crystal, Can Provide Subharmonic Frequencies," *Electronic Design,* November 8, 1974.

Carter, J. P. "Astable Operation of IC Timers Can Be Improved," *EDN,* June 20, 1973, p. 83.

Cicchiello, F. N. "IC Timer Yields 50% Duty Cycle," *Electronics,* May 13, 1976, p. 95.

Hilsher, R. W. "Constant Period With Variable Duty Cycle Obtained From 555 With Single Control," *Electronic Design,* July 5, 1975, p. 72.

Hofheimer, R. "One Extra Resistor Gives 555 Timer 50% Duty Cycle," *EDN,* March 5, 1974, p. 74.

James, T. W. "Single Diode Extends Duty-Cycle Range of Astable Circuit Built With Timer IC," *Electronic Design,* March 5, 1973, p. 80.

Klinger, A. R. "Single Part Minimizes Differences in Monostable and Astable Periods of 555," *Electronic Design,* July 19, 1974, p. 110.

Klinger, A. R. "Getting Extra Control Over Output Periods of IC Timer," *Electronics,* September 19, 1974, p. 112.

Klinger, A. R. "Generator's Duty Cycle Stays Constant Under Load," *Electronics,* November 28, 1974, p. 111.

Robbins, M. S. "IC Timer's Duty Cycle Can Stretch Over 99%," *Electronics,* June 21, 1973, p. 129.

POWER-SUPPLY CIRCUITS

Black, S. L. "555 as Switching Regulator Supplies Negative Voltage," *Electronics,* June 21, 1973, p. 130.

Bottomley, G. "Automatic Charger for Nickel-Cadmium Batteries," *Electronic Engineering,* December, 1975, p. 19.

Chetty, P. R. K. "IC Timers Control DC-DC Converters," *Electronics,* November 13, 1975, p. 121.

Chetty, P. R. K. "Put a 555 Timer in Your Next Switching Regulator Design," *EDN,* January 5, 1976, p. 72.

Domiciano, P. "Inverter Uses Ferrite Transformer to Eliminate Cross Conduction," *Electronic Design,* October 25, 1975, p. 130; see also "Comment Letter," *Electronic Design,* March 1, 1976, p. 7.

Durgavich, T. "Compact DC-DC Converter Yields ±15 V From +5 V," *Electronics,* June 21, 1975, p. 103.

Gartner, T. "IC Timer and Voltage Doubler Form DC-DC Converter," *Electronics,* August 22, 1974, p. 101.

Graham, G. A. "Low-Power DC-DC Converter," *Ham Radio,* March, 1975, p. 54.

Johnson, K. R. "High-Voltage Power Supply From 5-V Source Regulated by Timer Feedback Circuit," *Electronic Design,* April 1, 1975, p. 132.

Kranz, P. "A Simple Battery Charger for Gel Cells Detects Full Charge and Switches to Float," *Electronic Design,* July 19, 1976, p. 120.

McGowan, E. J. "IC Timer Automatically Monitors Battery Voltage," *Electronics,* June 21, 1973, p. 130.

Roll, I., M. Stienton, and D. Lucas. "Improvement to Automatic NiCd Battery Charger," *Electronic Engineering,* March, 1976, p. 17.

Solomon, R., and R. Broadway. "DC-to-DC Converter Uses IC Timer," *EDN,* September 5, 1973, p. 87.

Strange, M. "IC Timer Makes Transformerless Power Converter," *EDN,* December 20, 1973, p. 81.

MEASUREMENTS

Berlin, H. M. "555 Timer Tags Waveforms in Multiple Scope Display," *Electronics,* April 29, 1976, p. 114.

Blackburn, J. A. "Winking LED Notes Null for IC-Timer Resistance Bridge," *Electronics,* March 21, 1974, p. 100.

Blair, D. G. "Timer Chip Becomes Meter That Detects Capacitance Changes of 1 Part in 10^6," *Electronic Design,* March 1, 1976, p. 70.

DeKold, D. "IC Timer Converts Temperature to Frequency," *Electronics,* June 21, 1973, p. 131.

Ellison, J. H. "Universal L, C, R Bridge," *Ham Radio,* April, 1976, p. 54.

Graf, C. R. "Audio Continuity Tester Indicates Resistance Values," *Electronics,* April 1, 1976, p. 104.

Hall, C. "Direct-Reading Capacitance Meter," *Ham Radio,* April, 1975, p. 32.

Herring, L. W. "Timer ICs and LEDs Form Cable Tester," *Electronics,* May 10, 1973, p. 115.

Hinkle, F. E. "Overrange Indicator Can Enhance Frequency Meter," *Electronics,* April 17, 1975, p. 147.

Horton, R. "555 Timer Makes Simple Capacitance Meter," *EDN,* November 5, 1973, p. 81.

Klinger, A. R. "Logic Probe Built With IC Timer Is Compatible With TTL, HTL, and CMOS," *Electronic Design,* June 7, 1976, p. 156.
Mangieri, A. A. "The IC Photo Tachometer," *Popular Electronics,* August, 1974, p. 54.
Megirian, R. "Digital Capacitance Meter," *Ham Radio,* February, 1974, p. 20.
Mims, F. M. "LED Bargraph Readouts," *Popular Electronics,* September, 1976, p. 74.
Pepper, C. S. "Chopping Mode Improves Multiple-Trace Display," *Electronics,* October 14, 1976, p. 101.
Predescu, J. "Tester Built for Less Than $10 Gives GO/NO GO Check of Timer ICs," *Electronic Design,* May 24, 1974, p. 106.
Tandon, V. B. "Circuit Converts Single-Trace Scope to Dual-Trace Display for Logic Signals," *Electronic Design,* April 12, 1975, p. 80.

FUNCTION GENERATORS

Cicchiello, F. N. "Timer IC Stabilizes Sawtooth Generator," *Electronics,* March 18, 1976, p. 107.
Garland, M. M. "Simple Digital Waveform Synthesizer," *American Journal of Physics,* July, 1976, p. 710.
Gualtieri, D. M. "Triangular Waves From 555 Have Adjustable Symmetry," *Electronics,* January 8, 1976, p. 111.
Jung, W. G. "A Power Ramp Generator Delivers an Equally Adjustable 1-A Output," *Electronic Design,* March 1, 1976, p. 66.
Jung, W. G. "Build a Function Generator With a 555 Timer," *EDN,* October 5, 1976, p. 110.
Lloyd, M. A. "Function Generator With a Wide Frequency Range," *Electronic Engineering,* March, 1976, p. 23.
McClellan, A. "Current Source and 555 Timer Make Linear V-to-F Converter," *Electronics,* June 10, 1976, p. 108.
Penttinen, A. "Pulse Generator With Linearly Changing Frequency," *Electronic Engineering,* April, 1976, p. 25.
Reiter, T. "Frequency-to-Voltage Circuit," *Popular Electronics,* August, 1976, p. 75.
Tenny, R. "Linear VCO Made From a 555 Timer," *Electronic Design,* October 11, 1975, p. 96.

CONTROL

Bainter, J. R. "Dual-555 Timer Circuit Restarts Microprocessor," *Electronics,* March 18, 1976, p. 106.
Bockstahler, R. W. "Bistable Action of 555 Varies With Manufacturer," *Electronics,* February 19, 1976, p. 131.

Dekold, D. "IC Timer Plus Thermistor Can Control Temperature," *Electronics,* June 21, 1973, p. 128.

Dogra, S. "Operate a 555 Timer on a ±15-V Supply and Deliver Op-Amp Compatible Signals," *Electronic Design,* January 18, 1975, p. 76.

Gardner, M. R. "Line Drivers Made from 555 Timers Provide Inverted or Noninverted Outputs," *Electronic Design,* January 19, 1976, p. 86.

Gerek, F. "Potentiometer and Timer Control Up/Down Counter," *Electronics,* May 13, 1976, p. 94.

Graf, C. R. "Build a Light Sensitive Audio Oscillator," *EDN,* August 5, 1976, p. 83.

Gregson, P. H., and W. P. Lonc. "Light-Operated Millisecond Timers," *American Journal of Physics,* August, 1976, p. 803.

Heater, J. C. "Monolithic Timer Makes Convenient Touch Switch," *EDN,* December 1, 1972, p. 55.

Hinkle, F. E., and J. Edrington. "Timer IC and Photocell Can Vary LED Brightness," *Electronics,* December 26, 1974, p. 105.

Kraus, K. "Timer IC Paces Analog Divider," *Electronics,* August 5, 1976, p. 112.

Lewis, G. R. "Low-Cost Temperature Controller Built With Timer Circuit," *Electronic Design,* August 16, 1975, p. 82.

Locher, R. "IC Timer Gates High-Frequency SCR Circuit," *EDN,* January 5, 1974.

McDermott, R. M. "Oscilloscope Triggered Sweep: Another Job for IC Timer," *Electronics,* October 11, 1973, p. 125.

Mims, F. M. "TTL Sequence Generator," *Popular Electronics,* February, 1976, p. 101.

Murugesan, S. "Create a Versatile Logic Family With 555 Timers," *EDN,* September 5, 1976, p. 108.

Pate, J. G. "IC Timer Can Function a Low-Cost Line Receiver," *Electronics,* June 21, 1973, p. 132.

Potton, A. "Low-Cost Shaft Speed Monitor," *Electronic Engineering,* July, 1973, p. 49.

Reiter, T. "Light-Controlled Switch," *Popular Electronics,* March, 1976, p. 82.

Sarpangal, S. "IC Timer Drives Electric Fuel Pump," *Electronics,* November 25, 1976, p. 131.

Schlitt, G. "Monolithic Timer Generates 2-Phase Clock Pulses," *EDN,* August 1, 1972, p. 57.

Schulein, J. M. "Microprocessor Converts Pot Position to Digits," *Electronics,* March 4, 1976, p. 123.

Srinivasan, M. P. "Special-Purpose Pulse-Width Modulator Produces an Output of Same Polarity as Input," *Electronic Design,* September 27, 1976, p. 100.

Wellington, K. J. "Continuous Monitor for Seven-Segment Displays," *Electronics,* April 18, 1974, p. 118.

AUTOMOBILE AND HOME

Andre, R. "Low-Cost Stormcaster," *Popular Electronics,* August, 1974, p. 94.

Baxes, G. "Digital Fuel Gauge," *Popular Electronics,* December, 1976, p. 59.

Fox, T. R. "Measure the Wind," *Electronics Hobbyist,* Spring/Summer, 1976, p. 31.

Fusar, T. J. "IC Timer Makes Economical Automobile Voltage Regulator," *Electronics,* February 21, 1974, p. 100.

Galluzzi, P. "Circuit Provides Auto-Wiper Cycling, With One to 20 Seconds Between Sweeps," *Electronic Design,* December 20, 1974, p. 108.

Garner, L. "Solid State," *Popular Electronics,* May, 1974, p. 82.

Harvey, M. L. "Pair of IC Timers Sounds Auto Burglar Alarm," *Electronics,* June 21, 1973, p. 131.

Hilker, M. D. "Build a Digital Marine/Auto Tachometer," *Popular Electronics,* June, 1975, p. 40.

Kellem, C. S. "Slow-Sweep Wiper Control," *Popular Electronics,* April, 1975, p. 68.

Lewart, C. R. "Mobile Gas Alarm," *Elementary Electronics,* November/December, 1974, p. 31.

Lloyd, R. W. "Severe Weather Warning Alerter," *Popular Electronics,* May, 1976, p. 44.

Lo, C. C. "CD Ignition System Provides Low Engine Emissions," *EDN,* May 20, 1976, p. 94.

McVeigh, J. "Windshield Wiper Delay," *Popular Electronics,* November, 1976, p. 32.

Morgan, L. G. "Electronic Ignition System Uses Standard Components," *Electronic Design,* November 22, 1974, p. 198.

Redmile, B. D. "Tail-Biting One-Shot Keeps Car-Door Light On," *Electronics,* July 8, 1976, p. 92.

Williamson, T. A. "Alternately Flashing Taillights," *Popular Electronics,* March, 1975, p. 42.

Wyland, J., and E. R. Hnatek. "Unconventional Uses for IC Timers," *Electronic Design,* June 7, 1973, p. 84.

GAMES

Davies, J. R. "Personal Timing Tester," *Popular Electronics,* December, 1975, p. 77.

Frostholm, R. C., and R. Lundegard. "Tug-of-War," *Popular Electronics,* February, 1975, p. 43.
Nordquist, U. "Electronic Coin-Flipper," *Popular Electronics,* November, 1976, p. 91.
Pyska, M. "Random 4-Digit Number Generator," *Popular Electronics,* September, 1976, p. 100.

TELEPHONE CIRCUITS

Black, S. L. "Get Square-Wave Tone Bursts With a Single Timer IC," *Electronic Design,* September 1, 1973, p. 148.
Dugan, K. "Making Music With IC Timers," *Electronics,* April 18, 1974, p. 106.
Dugan, K. "Ringer Enables Telephone to Play Simple Tune," *Electronics,* May 15, 1975, p. 115.
Herring, L. W. "Generating Tone Bursts With Only Two IC Timers," *Electronics,* May 30, 1974, p. 107.
Kraengel, W. D. "Optically Coupled Ringer Doesn't Load Phone Line," *Electronics,* February 20, 1975, p. 92.

HAM AND CB RADIO

Berlin, H. M. "A Transceiver Actuated Time-Out Warning Indicator for FM Repeater Users," *Ham Radio,* June, 1976, p. 62.
Blakeslee, D. "Time—IC Controlled," *QST,* June, 1972, p. 37.
Buswell, J. "Simple Integrated-Circuit Electronic Keyers," *Ham Radio,* March, 1973, p. 38.
Conklin, B. "Identification Timer," *Ham Radio,* November, 1974, p. 60.
Klinert, C. "Repeater Keying Line Control," *73,* February, 1973, p. 60.
McVeigh, J. J. "Build the Five-Minute 'On' One-Minute 'Off' Timer," *Popular Electronics,* April, 1976, p. 60.
Vancura, W. J. "A Simple Electronic Keyer for Sending Morse Code," *Popular Electronics,* August, 1976, p. 44.
Vordenbaum, H. "Automatic Reset Timer," *Ham Radio,* October, 1974, p. 50.
Wooten, W. L. "A Code Practice Oscillator for the Beginner," *QST,* November, 1972, p. 38.

PHOTOGRAPHY

Giannelli, J. "Automatic Photo Enlarger Controller," *Popular Electronics,* April, 1974, p. 51.

Mangieri, A. A. "Photo Timer," *Electronics Hobbyist,* Spring/Summer, 1976, p. 67.
Marchant, R. "IC Photo Development Timer," *Popular Electronics,* October, 1973, p. 70.

Index

Index

TO THE READER

This book is one of an expanding series of books that will cover the field of basic electronics and digital electronics from basic gates and flip-flops through microcomputers and digital telecommunications. We are attempting to develop a mailing list of individuals who would like to receive information on the series. We would be delighted to add your name to it if you would fill in the information below and mail this sheet to us. Thanks.

1. I have the following books:

____

____

____

____

____

____

____

____

____

____

____

____

2. My occupation is: ☐ student ☐ teacher, instructor ☐ hobbyist

☐ housewife ☐ scientist, engineer, doctor, etc. ☐ businessman

☐ Other: ____________________

Name (print): ______________________________

Address ______________________________

City ____________________ State ____________

Zip Code ________________

Mail to:

Books
P.O. Box 715
Blacksburg, Virginia 24060